D1005954

A NEW HUMAN

Smithsonian Books

The Startling Discovery and Strange Story of the "Hobbits" of Flores, Indonesia

Mike Morwood
and
Penny van Oosterzee

DOUGLAS COLLEGE LIBRARY

A NEW HUMAN. Copyright © 2007 by Mike Morwood and Penny van Oosterzee. All rights reserved. Printed in the United States of America. No part of this book may be used or reproduced in any manner whatsoever without written permission except in the case of brief quotations embodied in critical articles and reviews. For information address HarperCollins Publishers, 10 East 53rd Street, New York, NY 10022.

HarperCollins books may be purchased for educational, business, or sales promotional use. For information please write: Special Markets Department, HarperCollins Publishers, 10 East 53rd Street, New York, NY 10022.

First Smithsonian Books edition published 2007.

Designed by Stephanie Huntwork

Library of Congress Cataloging-in-Publication Data

Morwood, M. J. (Mike J.)
A new human: the startling discovery and strange story of the "hobbits" of Flores, Indonesia / Mike Morwood and Penny van Oosterzee. — 1st Smithsonian Books ed.
p. cm.
ISBN: 978-0-06-089908-0
ISBN-10: 0-06-089908-5
1. Fossil hominids—Indonesia—Flores Island. 2. Fossil hominids—Australia—Kimberley (W.A.) 3. Pygmies—Indonesia—Flores Island. 4. Pygmies—Australia—Kimberley (W.A.) 5. Human beings—Migrations. 6. Excavations (Archaeology)—Indonesia—Flores Island. 7. Excavations (Archaeology)—Australia—Kimberley (W.A.) 8. Human remains (Archaeology)—Indonesia—Flores Island. 9. Human remains (Archaeology)—Australia—Kimberley (W.A.) 10. Flores Island (Indonesia)—Antiquities. 11. Kimberley (W.A.)—Antiquities. I. Van Oosterzee, Penny. II. Title.

GN730.32.I5M67 2007
994.01—dc22

2006052267

02 03 04 05 06 WBC/RRD 9 8 7 6 5 4 3 2 1

CONTENTS

PROLOGUE

Jack tossed a match into the clump of spinifex, which exploded into flame. A few more matches in a few more clumps and there was a wall of fire up to five meters high and spreading fast. The accompanying roar made it difficult to hear anything else, and the acrid smoke made it difficult to see, so we hastily retreated to our boat beached among the mangroves. As we motored to our camp across the inlet, I looked back to see a gigantic pall of smoke reaching hundreds of meters into the sky with Jack at the back of the boat silhouetted against the inferno. Jack Karadada, a Wunambal Aboriginal elder and one of the traditional owners for this remote stretch of the Kimberley coastline in the northwest corner of Australia, had taken the opportunity of our visit to clean up his country, which by traditional custom should be burned regularly. The fire would burn for weeks across this vast, rugged, empty landscape.

Until Jack's cleaning up the country brought our work to a halt, my colleague Doug Hobbs and I had been recording a major Indonesian site for processing of trepang, or sea cucumbers—a group of animals related to sea stars and sea urchins that the Chinese value as a delicacy and an aphrodisiac. At this site, we had found evidence for processing on an industrial scale. There were 18 lines of stone fireplaces that had supported cauldrons for boiling trepang, while scattered on the ground all around were pieces of pottery from Java, Sulawesi and China, left by Macassan, Buginese and Bajau fishermen.

There are hundreds of such processing sites around the coasts of the Kimberley and Arnhem Land—known to Indonesian fishermen as Marege and Kaju Djawa, respectively. Large-scale Indonesian visits to northern Australia to collect trepang began in historical times, and the actual date for the commencement of the industry is known. Meticulously kept records of the Dutch East Indies Company show that from 1700 CE, the fishermen came sailing on the northwest monsoon winds to northern Australia for trepang, which they collected in shallow coastal waters, then brought ashore to be gutted, boiled and smoked in prefabricated smokehouses. The fishermen stayed until the winds swung around, when they sailed back on the southeast trades, laden with trepang for the Chinese market.

An Indonesian trepang-processing site drawn around 1843–1844. Remains of such Asian sites are found along the coast of the Kimberly, Arnhem Land and the Gulf of Carpentaria. (*THE QUEEN*, FEBRUARY 1862)

That was not the first time Asians had come to Australia. Archaeological evidence suggests much earlier visits by seafarers, before the trepang trade and long before historical records were kept. The appearance of the dog in Australia about 4,000 years ago, for instance, must have resulted from Asian contact—similarly the appearance of the pig in New Guinea about the same time. Going back further still, the initial colonization of Australia and New Guinea by modern humans at least 50,000 years ago is another indisputable example of outside contact: the fossil record makes it clear that humans did not originate there.

The complex rock art styles and languages found in northwest Australia suggest cultural exchange and population movement over a long time between Asia, New Guinea and Australia. The exquisite Bradshaw and stylistically related Dynamic Figure rock paintings, for instance, which are more than 20,000 years old, have a degree of anatomical detail, composition and movement seldom matched in any other Australian Aboriginal art tradition, but are very similar to rock paintings found in Borneo that are more than 10,000 years old. Bradshaws, Dynamics and other Complex Figurative art styles, such as Wandjina and X-ray paintings, are only found in the Kimberley, Arnhem Land and adjacent regions of northwest Australia.

The sophistication and complexity of the art is matched by the diversity of languages: there are 18 Australian Aboriginal language families, 17 of which are found only in the northwest. The hundreds of languages found throughout the rest of the continent all belong to just one language family, Pama-Nyungan.

For 20 years, as a lecturer at the University of New England (UNE) in Armidale, New South Wales, I had been doing research on Australian Aboriginal archaeology, progressively working farther north, then west, with regional projects in Southeast Queensland,

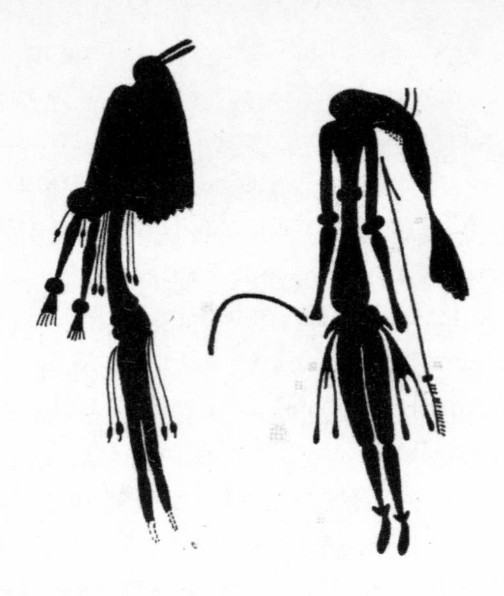

Bradshaw rock paintings in the Kimberly region of northwest Australia are more than 20,000 years old. Only rock art styles in the Kimberly, Arnhem Land and adjacent regions of the northwest have this degree of figurative detail. (CREDIT: MIKE MORWOOD, AFTER WALSH, 2000)

the Central Queensland Highlands, the North Queensland Highlands, Cape York Peninsula, and then the Kimberley—the latter being one of the likely beachheads for the first people to reach these shores.

When those first migrants crossed from continental Asia to New Guinea, mainland Australia and Tasmania, connected as one "Greater Australia" throughout most of the evolution of modern humans, it was actually possible to see from one island to the next all the way. There were, in fact, no fewer than five such routes across the scattering of 13,000 islands that make up the Indonesian archipelago. Three of these routes snaked across Sulawesi, streaming with the currents, through the kingdom of a thousand islands that is the Moluccas, with two landfalls at the western tip of New Guinea, and one to the south on what today is the island of Aru, but was then part of the Greater Australian coastline. The other two routes involved island hopping, against the currents, along the Nusa Tenggara islands from Lombok to Flores, and then leaping 100 kilometers either to the Kimberley via Timor or to the Aru section of the Greater Australian mainland via Tanimbar. All these routes involved an initial crossing of the deep, enigmatic Wallace Line, a moat separating mainland Asia and the world of islands between Asia and Australia. The ancestors of the First Australians made this crossing, and almost certainly came via Indonesia. But who were they exactly? How did they come? And why?

Doug, Jack and I were camped on the Anjo Peninsula, a narrow spit of land projecting into the Timor Sea, where we could watch the sun descend into the sea to the west while a full moon rose out of the same sea to the east. A good place to think about what you are doing with your life and why. Doug and I discussed starting an archaeological research project in Indonesia to search for the origins of the

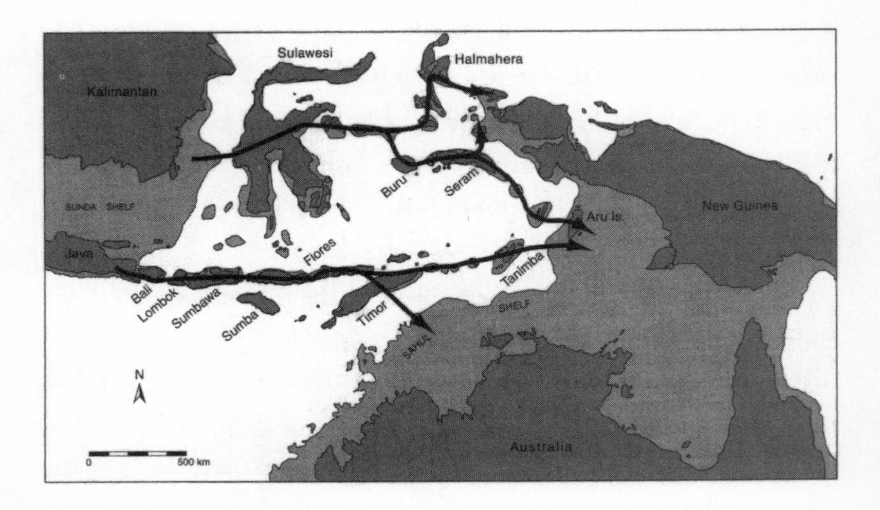

Five possible migration routes of the initial human colonization of Australia. For each of these routes it is possible to see evidence for the next island. (CREDIT: MIKE MORWOOD, AFTER BIRDSELL, 1977)

First Australians. There seemed to be all sorts of practical objections—others had tried and not got very far in the complex world of Indonesian bureaucracy and politics—but we decided to give it a go anyway.

That is how it all started.

In 2003, some eight years after my Kimberley musings and as a direct result, Thomas Sutikna, Wahyu Saptomo, Jatmiko, Rokus Awe Due and Sri Wasisto, five researchers from the Indonesian Research National Centre for Archaeology, found an almost complete skeleton of a tiny adult woman who died 18,000 years ago at Liang Bua, a large limestone cave on Flores. The skeleton, known affectionately as "Hobbit," is now the type specimen for a new human species, *Homo floresiensis*.

This is the story of the events leading up to the discovery of *Homo floresiensis*—from my first tentative visit to Indonesia; to research on early human sites in the Soa Basin of central Flores, which provided the disciplinary credence and the network of colleagues needed to tackle other big questions in Southeast Asian archaeology; to the planning of the collaborative Indonesian-Australian research program under which the Liang Bua excavations were carried out. And finally to the discovery of the tiny, new human species, how it compares with other early hominid finds, and the paradigm-shattering implications.

Homo floresiensis challenges us because she is so unexpected, because she does not fit with many preconceptions about how humans evolved and behaved, and what they should look like. Taken in the context of what happens to other land animals when they become established on isolated islands, she is, however, exactly what we might expect. Some find this possibility not to their liking and have challenged it, which in turn has led to a sometimes bizarre series of twists and turns in Hobbit's postexcavation history, all with astonishingly similar precedents in the history of paleoanthropology.

These are not just issues and events of interest to researchers, academics or philosophers, as the popular reaction to *Homo floresiensis* has shown over the past two years. Such a level of interest has to be good for paleoanthropology and archaeology even if some negative aspects of her story are publicly aired for the first time.

Although Penny van Oosterzee and I have written this book from the perspective of Mike Morwood, with all the weaknesses and biases that that entails, we would like to emphasize that the findings reported here result from the hard work and expertise of a large team of researchers, from a number of disciplines and institutions, in Indonesia, Australia, America, Canada and Britain.

—*Mike Morwood*

1

In the Footsteps of the Father

Getting a foothold in Indonesia was a daunting prospect. A different country, a different culture, a different language. One way to ease the process would be to get the assistance and advice of people with prior experience. So in 1995, having decided while in the Kimberley to try looking for the Asian origins of the First Australians, I wrote to people who were already doing archaeological research in the region. No reply. Time to jump in at the deep end; to go to Jakarta, the center of politics, power and wealth in Indonesia, to meet people and set up collaborate research projects.

Jakarta is hot, humid and chaotic: a sprawling capital of 10 million people mainly living in crowded residential areas, with intermittent clusters of high-rise buildings and all connected by traffic jams. Clutching an *Indonesia* Lonely Planet guidebook, I made my way to the National Research Centre for Archaeology (ARKENAS) to meet "the father of archaeology in Indonesia," Professor Raden Pandji Soejono, then aged 66. Soejono, whose honorific title "Raden" means "Prince," is an aristocrat whose ancestors ruled the last great

Hindu kingdom in Java, the Majapahit, in the 14th century. He is the son of a hero in the struggle for Indonesian independence, Raden Pandji Soeroso, who had become a cabinet minister in the Soekarno government. Soejono himself became famous when, as a young man during the Japanese occupation in World War II, he climbed a flagpole to rip down a Japanese flag, but was persuaded otherwise by a Japanese soldier who took aim with a rifle but did not fire.

Until his retirement in 1987, Soejono was director of ARKENAS, and he still occupied two adjacent large offices in the building. Symbolically retaining the premiere parking space for his car right next to the front steps, he had a major influence on decisions and policies made by subsequent directors. A man of medium build, with distinguished gray hair combed back and thinning, Soejono wears well-cut suits, and is fluent in a number of languages including English and Dutch. Over his long career, he had excavated at sites in many parts of Indonesia, and had been president of the Indo-Pacific Prehistory Association. He had the knowledge and contacts to get a cooperative project established. But although he was very polite and interested in my Australian research and ideas for future research in Indonesia, nothing quite jelled. I wasn't surprised. This had also been the experience of two archaeologist colleagues from the Australian National University, Rhys Jones and Alan Thorne, who had visited Indonesia many times, had many polite conversations with Soejono and even looked at sites in Timor. All to no avail. I later learned that one of the sites they had hoped to work at collaboratively was a large limestone cave in western Flores where Soejono had been working since 1978. Its name was Liang Bua.

My experience with Dr. Fachroel Aziz, a paleontologist with the Geological Research and Development Centre (GRDC) in

Bandung, was very different. In fact, when I phoned him from Jakarta, he was immediately enthusiastic and wanted me to come straight there to examine some stone fragments, possibly stone artifacts, that he and some Dutch colleagues—Paul Sondaar, John de Vos and Gert van den Bergh—had excavated the previous year from Mata Menge, an open site in the Soa Basin of central Flores. The fragments had been found in the same sandstone layers as the fossilized remains of *Stegodon,* an extinct group of animals, related to mastodons, mammoths and elephants, and once common across Asia.

Aziz, from a humble background in northern Sumatra, was self-made, much more entrepreneurial and less formal than the aristocratic Soejono—and they did not get on. That was probably the main reason that Aziz did not seek the advice of ARKENAS archaeologists about the Mata Menge stones, but preferred to deal with a foreign archaeologist—a newcomer to Indonesia, who was politically naïve, had no prior affiliation with any Indonesian institution, and had not been mentored by Soejono. I went by train to Bandung the same day to take up Aziz's offer, and immediately set to work studying the stone pieces from Mata Menge laid out on the desk in his office.

Aziz and his colleagues were not the first to excavate at Mata Menge. The seminal work at the site had been done in 1963 by Father Theodor Verhoeven, a Catholic priest based in Flores and with a passion for archaeology, who had studied Pompeii for his master's degree in classical history at the University of Leiden. At Mata Menge and another nearby fossil site, Boa Lesa, he found stone artifacts, including flake tools, chopping tools, and hand axes, together with *Stegodon* remains in sandstone layers, sandwiched between thick layers of volcanic ash. Quite logically, Verhoeven concluded that early humans and *Stegodon* coexisted on Flores. But then he claimed something seemingly preposterous: because *Stegodon* and *Homo erectus* were known to have lived in Java about 750,000 years ago, he concluded that the stone tools at Mata Menge were of similar age, and that *Homo erectus* had somehow reached Flores.

Following further excavations in the Soa Basin with another priest, Father Johannes Maringer, in 1970 Verhoeven presented the evidence for his claims in a number of papers in the journal *Anthropos*. His evidence was ignored by the archaeological establishment because of doubts about his identification of stone artifacts, the possibility that any actual stone tools might have become mixed up with much older fossils, and the fact that no one knew when *Stegodon* had lived on Flores. Verhoeven also published his findings in German, which made it even easier for his detractors simply to ignore them.

Father Theodor Verhoeven excavating at Mata Menge in the Soa Basin in 1963. Here he recovered the first evidence for stone artifacts in the same deposits as Stegodon *fossils. But the archaeological establishment refused to accept his findings or conclusions.* (PHOTO: THEODOR VERHOEVEN)

It didn't help that Verhoeven had fallen out with H. R. van Heekeren and Dirk Hooijer, two of the major figures in Indonesian archaeology and paleontology at the time. In 1967, van Heekeren, author of an influential book, *The Stone Age of Indonesia,* and mentor to Soejono, had arranged with Verhoeven to visit archaeological sites in Flores, but had come earlier than agreed and at a time when Verhoeven happened to be away. So van Heekeren visited a number of Flores sites, including Liang Bua, by himself before Verhoeven returned. This initiated a major falling-out between the two men. A similar falling-out occurred when Hooijer reneged on an agreement to return some Flores *Stegodon* fossils that Verhoeven had sent to him in the Netherlands. Furthermore, both van Heekeren and Hooijer regarded Verhoeven as an amateur, and resented his cooperation with Ralph von Koenigswald, an ambitious paleontologist with the Geological Survey of the Dutch East Indies whom they disliked. Such personal animosities and loyalties may have been a major reason for Verhoeven's evidence and claims being long ignored. A pity because the claims for early humans on Flores have profound implications, particularly because Flores is an island, and has been for millions of years.

Located almost exactly halfway between the Asian and Australian continental areas, Flores is right on the geographical, cultural and linguistic boundary between Asia and Australia-Melanesia, and a possible route for initial colonization of Greater Australia by modern humans. But even at low sea levels, at least two sea crossings are needed to reach the island. The first of these deepwater sea barriers is a 25-kilometer strait between the islands of Bali and Lombok; the second is a 9-kilometer strait between Sumbawa and Flores. Up until recently, it was assumed that only modern humans had the required intellectual, linguistic and technological capacity to make sea crossings. The Flores evidence demonstrated

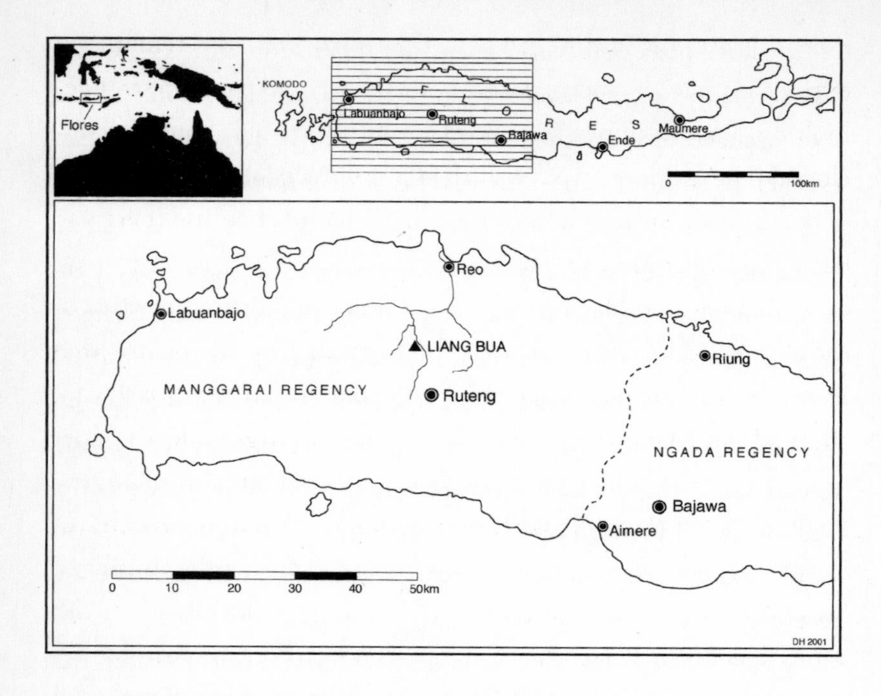

Map showing the location of Flores and Liang Bua. The Soa Basin occurs in the Ngadha regency immediately to the east of Bajawa. (CREDIT: MIKE MORWOOD)

that this assumption might not be correct, which had mind-blowing implications. It made me realize that early humans must have been much smarter than most people think. People tend to consistently underestimate the abilities and achievements of our ancestors.

It would take another 30 years before further research was undertaken at Mata Menge, when Aziz and Paul Sondaar decided to investigate Verhoeven's claims. Sondaar, a lecturer at Utrecht University in the Netherlands, who studied paleontology under Ralph von Koenigswald and was mentor to Aziz, had a long-standing research interest in the evolution of animals on islands. In 1984, he invited Aziz to his excavations at Corbeddu Cave on the Mediterranean island of Sardinia, where they recovered the remains of extinct animals, including large-bodied deer, and part of a strange-looking, human upper jaw about 18,000 years old. Sondaar concluded that the deer on Sardinia had not reduced in size over time, as those on the nearby island of Crete had done, because there had been a predator present—humans.

A long-retired and now elderly Verhoeven also spent time at those Corbeddu Cave excavations. Intrigued by his claims that *Homo erectus* had reached Flores, Sondaar visited the Soa Basin, checked out Mata Menge and also found a stratigraphically older site, Tangi Talo, which contained the fossilized remains of pygmy *Stegodon*, giant tortoise and Komodo dragon. He immediately realized that here was a real opportunity to examine the impact of early humans on a pristine island. More specifically, Sondaar wanted to know if the arrival of humans had been responsible for the extinction of pygmy *Stegodon* and giant tortoise on Flores, and whether the large-bodied *Stegodon* that subsequently recolonized the island had not reduced in size over time precisely because of the presence of human hunters.

During a field trip to Flores in 1994, Sondaar, Aziz and their team undertook two excavations—the first at Mata Menge, where they recovered the remains of large-bodied *Stegodon* fossils in the same strata as possible stone artifacts, and the second at Tangi Talo, which had abundant evidence for pygmy *Stegodon*, giant tortoise and Komodo dragon, but apparently no stone artifacts. They also took a series of rock samples for paleomagnetic dating of both sites.

The paleomagnetic dating technique is based on the wanderings of the magnetic North and South Poles. Every now and then, there are complete reversals of the magnetic field as a result of changes in current flow at the core of the planet, and this is recorded in the magnetic properties of rocks formed at the time.

Such reversals happen roughly every million years. The present period is called the Brunhes and it began 780,000 years ago. The preceding period, the Matuyama, began 2.5 million years ago and had the opposite polarity. The difficulty of working out which reversal is which is compounded by the fact that in between main reversals there have been short-term reversals. Because there have been many changes in polarity, the results of paleomag studies can be ambiguous, and other evidence is usually needed to decide which fluctuations the rock has captured.

The Indonesian-Dutch team concluded on the basis of the paleomag results that Tangi Talo, with pygmy *Stegodon*, giant tortoise and no stone artifacts, was about 900,000 years old, while the site of Mata Menge, with large-bodied *Stegodon* and stone artifacts, was about 750,000. The smoking gun of the stone tools clearly pointed to the arrival of humans as the cause of this faunal turnover—an extraordinary finding because it would be the first record of an early species of human causing extinctions, a trait thought to have been the exclusive province of us modern humans. Unfortunately, these conclusions were published in an unrefereed, fairly obscure venue, and because their team did not include researchers with stone artifact expertise, the archaeological community remained dubious about the identification of "stone artifacts" at Mata Menge.

Back in Bandung, my analysis of the stone pieces excavated from Mata Menge showed that yes, most were artifacts. Aziz was delighted by this result, which we published with Sondaar, de Vos and van den Bergh. It put on record the definite existence of stone artifacts at the site. Verhoeven had been vindicated—humans had been present on Flores at the same time as *Stegodon*. This was an important first step in getting wider acceptance of the claims for early human colonization of Flores. The next step was to better determine the age of the site, examine its stratigraphy—the all important layering of the sediments that, like a book, tells the story of the site—and ascertain how the fossils and artifacts came to be there. Our aim was to publish the evidence in an international refereed journal, giving the claim much greater credibility and exposure than it ever achieved previously. This was exciting. I was finally tackling the big questions in the archaeology of the region. This is what I had hoped for when camped on the Anjo Peninsula back in the Kimberley. This is why I had come to Indonesia.

The following year, 1996, Aziz and I made a brief trip to Flores to visit Mata Menge and collect this evidence. The trip did not start well. We had to overnight in Jakarta to catch an early morning flight to Kupang, capital of West Timor and the administrative center for islands east of Bali, to obtain research permissions from the departments of Education and Culture, Social Politics, and Police. We had allowed a good hour and a half to get to the airport in time for the flight, but during the night there was torrential rain with widespread flooding and major traffic jams. Traffic in Jakarta is always difficult, but even more so in those conditions. It seemed we would never make it to the airport in time. Aziz promised our driver a bonus if we made the flight, so he drove with determination, along footpaths, and the wrong way down flooded one-way streets. Getting to the airport, we raced inside just as the final boarding call was being made for the Kupang flight—and boarded. Any later and this story might have turned out very differently.

Downtown Kupang is busy, grubby, hot and sticky, like many harbor towns in the region. Inland from the commercial area, which hugs the coastal flat, are the dry, rocky limestone hills that sit above the sweeping panoramic bay of Kupang with its moored fleets of fishing boats—manned by Macassan, Buginese and Bajau crews. On the hill slopes with their cooling winds and views, there are residential areas and higher still are the grand office buildings of various government departments. We had to do the rounds over two days to get local government permits for our work. Fortunately, Aziz knew how the system worked and government employees would often deliver the letters of approval to our hotel after hours for a small commission.

My strongest memory of Kupang is of the *bemos,* the minivans that provide the main form of public transport there. Different-numbered bemos travel in prescribed routes that cover the whole city and outlying regions. It is a very efficient system, where you never have to wait for more than five minutes before the right-numbered bemo comes along. And they are easy to spot and hear. In Kupang, bemos try to look and sound as much like mobile discos as possible. Beneath the bench seats for passengers, there are usually two large speakers playing hard-rock music, with the bass turned right up, creating a "womper womper" sound that can be heard long before the vehicle comes in sight. As well, many bemos have revolving lights just above the passengers' heads, while the interior is plastered with garish pop posters. How the driver manages to see out the windscreen is a matter of conjecture, given the strips of flashing colored lights that obscure most of it. We were told that all this was for the benefit of the passengers, and that they preferred to patronize the loudest, most colorful vehicles. Looking at the other passengers during one ride—a woman with two young children, a middle-aged man and two old ladies—Aziz and I doubted this was so, and suspected it had more to do with the tastes of the macho young men who drive bemos. On behalf of all, we respectfully yelled for the volume to be turned down.

Having got our permits, we flew from Kupang to Ende on the central south coast of Flores, a small town surrounded by steep, rugged volcanic hills. The island, at 13,500 square kilometers, is the largest in the Nusa Tenggara, or Lesser Sunda, chain of islands that comprises dozens of volcanic and coral-reef islands strewn across the world's deepest seas, which plunge down seven kilometers. It is characterized by rugged volcanic mountains up to 2,400 meters high, deep canyons and gravel plains. The main range runs east–west the length of the island and sheds water to the north and south coasts.

On Flores you're invariably walking on shaky ground because your feet are planted directly over the greatest tectonic collision zone on Earth. The Indonesian archipelago, with its chain of volcanic cauldrons spread over 3,000 kilometers, is itself the surface expression of the collision between the Indian-Australian and Eurasian tectonic plates. The thinner oceanic Australian plate, moving northward at around eight centimeters a year, plunges steeply beneath the lighter continental Eurasian plate. The Australian plate doesn't slide smoothly underneath the Asian plate; instead it moves in jolts, buckling and carrying bits of Asia with it. These jolts are felt as earthquakes, which occur from near the surface, where the Australian plate begins its descent, to depths of 700 kilometers, where the plate is caught up in the slowly moving molten material of Earth's mantle.

As a result, Indonesia is a world leader in volcano statistics. At 76, it has the largest number of historically active volcanoes, with at least 132 active during the last 10,000 years. Indonesia's total of around 1,200 dated eruptions is only narrowly exceeded by Japan's of almost 1,300. The archipelago also has suffered the highest number of fatalities caused by eruptions and the accompanying mudflows, tsunamis, giant upwelling mounds of lava, known as domes, and pyroclastic flows. The devastating effects of recent eruptions, such as Krakatoa in the Sunda Strait between Sumatra and Java in 1883 and Tambora on Sumbawa in 1815, are well documented, but we can only infer what impacts earlier events had on human populations. The Toba eruption on Sumatra 74,000 years ago, for instance, was more than a hundred times bigger than the historic Krakatoa eruption, and put so much ash into the atmosphere that it resulted in a worldwide temperature decrease that lasted five years. Indonesia is a land of global-scale fireworks, and the island of Flores has seen more than its fair share.

From Ende we went by rental car to Bajawa, capital of the Ngadha regency, for another round of permits from the local offices of the departments of Education and Culture, Social Politics, and Police. The local Ngadha are dark-skinned and Melanesian in appearance, and are a highly distinctive ethnic group. Their oral history recounts how they moved into the area from the west and first settled around the sacred mountain, Inerie, about 30 generations, or 600 years, ago—with *nga dha* meaning "nomads" or "people who travel far."

Like many other proud hill peoples, the Ngadha are renowned warriors, who served as mercenaries in other people's wars and staved off Dutch conquest until 1907. Men carry long knives (parangs), which they use in clearing vegetation, slaughtering animals, woodworking and, occasionally, in disputes, on each other. In fact, during a later field season, one of our workers at Mata Menge, while ostensibly guarding the excavations, went over to a nearby field one afternoon and killed a man with whom he had had a long-running dispute over land inheritance. The victim had defense wounds to his forearms, a parang cut to the back of the neck as he turned to run, and had been finished off on the ground with a blow to the mouth that penetrated the brain. You now pass his grave on the way to Mata Menge. The murderer went on the run and avoided capture by the police for a few days, while the victim's relatives burned down the assailant's house and were only just dissuaded from killing his wife.

The Ngadha cultivate corn, coffee and rice, but given the rugged nature of their region, most of which is marginal for cultivation, they are much more pastoral than other ethnic groups like the Manggarai immediately to the west. Cattle, buffalo, horses, pigs and chickens represent wealth, since they are used as gifts, bridewealth, restitution, blood sacrifices and for payments of any kind.

They have their own distinctive languages, dress, art, music and traditional religious beliefs (*adat*) that underscore their Roman Catholicism. Ngadha symbols and ceremonies, therefore, involve a complex mixture of beliefs. For instance, traditional houses can have crucifixes and prints depicting the life of Christ, or the pope, adorning the walls in addition to pairs of carved wooden figures that represent male and female clan ancestors near the entrance. These *anadeo* figures, freestanding or shown together on horseback, serve to protect the household against spirits and ensure good luck, wealth and fertility.

During a house completion ceremony that I attended at a village called Wogo Baru, the Lord's Prayer was recited, the village was circled by elders bearing bundles of thatch, a gourd of the local alcoholic beverage—palm wine, or *moki*—and a chicken. Feverishly, the younger men then completed the roof of the house with bundles of thatch. Nine pigs slung from poles were carried in and laid on the ground, front legs tied behind. As soon as the roof was completed, young men with parangs split the heads of the squealing pigs, while others held buckets to collect the blood. In this way, the new house became "named": the occupants of the house were blessed with wealth and fertility, and the jawbones and tusks of the sacrificed pigs would be displayed permanently at the front of the house for all to see.

The arrangement of Ngadha villages also reflects traditional beliefs that emphasize maintaining good relations with ancestors. Traditional villages, such as Bena, typically have a central plaza, divided up into areas called *loka*, by stone walling and with large stone monuments, or megaliths, representing the various subclans, or *woe*, of the village occupants. Megaliths are often the graves of important ancestors who played a prominent role in the acquisition of clan lands. As well as symbols of clan identity and power, megaliths therefore serve as charters for land ownership and provide a ceremonial focus for clan activities. In some ceremonies, a buffalo is tethered by a long lead to a megalith and then hacked to death with parangs. The blood shed by the beast running around the *loka* ensures fertility. For the same reason, boxing matches are held in which both men have fiber cords with embedded pieces of glass wrapped around their fists to ensure that blood is shed during the bouts.

Knowing about the Soa Basin didn't prepare me for its physical presence. It is a geologically stunning expanse, stretching from horizon to horizon, largely unmarred by buildings or people—a volcanically encircled savanna, or grassland, with numerous small rounded hillocks, occasional deep gorges cut by the Ai Sissa River and its tributaries, and the perfectly conical Ambolobo volcano at the southeastern end.

The little puffs of white smoke issuing from Ambolobo and other volcanoes surrounding the Soa Basin looked harmless enough, but that can easily change, as it has many times in the past. The molten magma deep within the earth is like carbonated beverage in a bottle. As the Indian-Australian continental plate slides into earth beneath the Eurasian plate, it drags with it water, carbon dioxide, and minerals containing sodium and potassium from the oceanic sediment. They all act to lower the melting point of the magma, and increase its gas content. In a volcanic eruption the gas, like bubbles in champagne, explodes out of solution, shattering the magma into sand and grit-sized volcanic ash at the moment of eruption. Ash, with water, combines to form mud, which avalanches down the sides of the volcanoes, and rains from the sky. Pumice, cinders, blocks and molten lava bombs are the explosive products of gas and magma. Vulcanologists call these "pyroclastics"—literally, "fire fragments"—and the different types of pyroclasts, and the order in which they are produced, give volcanoes their characteristic shape. Alternations of volcanic ash and lava flows built the beautiful, steep, nearly perfect cones with concave slopes, called stratovolcanoes, like Ambolobo, looming beautifully ominous over the Soa Basin.

Another striking aspect is the fact that very few people live in the 1,000 square kilometers of savanna that constitute the Soa Basin. There is the occasional hut, as well as rice paddy systems where water is available, but the area is hot and arid, and generally the only people you see are groups of hunters on horseback, and men looking after herds of cattle, buffalo or horses.

From a geological point of view the Soa Basin is like a cake, layered with largely undisturbed sediments, each layer telling its own story. The base of the Soa Basin is an extremely hard jumble of volcanic rock, a breccia, called the Ola Kile Formation, which is overlain by a series of pumice-rich layers, known as tuffs, which are visible as thick white bands in the landscape. These were formed by pyroclastic flows, fast-moving clouds of volcanic ash and hot air. From about a million to 650,000 years ago, river outlets from the basin were periodically blocked to form freshwater lakes that were drained when the river cut its way through the surrounding rim of volcanoes again, returning the area to savanna. Herds of *Stegodon* wandered through the basin, feeding, fighting, and dying, leaving their remains to be scavenged by Komodo dragons and washed into channels and lakes. Deposits of airborne volcanic ash and materials eroded from the surrounding mountains then sealed in the remains of *Stegodon*, Komodo dragon, crocodiles, rats, birds, molluscs, and plant remains.

Over time, these deposits, now known as the Ola Bula Formation, accumulated to a depth of 120 meters. In turn, they were capped by the Gero Formation, made up of layers of limestone formed in shallow, freshwater lakes that periodically dried up. A million years ago the Soa Basin was located just above sea level, but it has continued to rise about a centimeter per century, pushed up by tectonic forces along with the rest of Flores. So sediments within the basin are now being eroded to expose deeply buried layers of ash, sandstones and the fossils they contain.

For the most part, those sedimentary layers were laid down horizontally, and they have remained so. As a result, the relative height of geological layers and sites across the Soa Basin can be used as an indicator of their relative age and show that a major change in local animals occurred. At the oldest known fossil site, Tangi Talo, there are remains of pygmy *Stegodon*, giant tortoise and Komodo dragon. Higher, and therefore younger, sites contain evidence for full-size *Stegodon*, Komodo dragon and rat.

On that first visit with Aziz in 1996, as well as subsequent visits to the area, we stayed in the large, sprawling house of the rajah's widow in Boawae, a small town at the base of the magnificent Ambolobo volcano on the southeastern edge of the Soa Basin. The rajah of Nagakeo, the "local king," had died about two years before and, as is custom in Flores, has a well-tended grave in front of the house. In another five years' time, there would be a major sacrifice of buffalo as part of the long-term ceremonies required to mark his passing. The rajah's father had been the one to first draw Verhoeven's attention to large bones eroding out of a site called Ola Bua. These turned out to be the fossilized bones of *Stegodon,* and the discovery initiated Verhoeven's research on the paleontology and archaeology of Flores.

We didn't have much money or time during our first visit in 1996, so our initial goals were limited. Aziz and I were determined to record the stratigraphy at Mata Menge and Tangi Talo in detail, to assess the dating opportunities, and to take samples for dating. In this way some of the shortcomings of the previous work at the sites would be addressed, and we would finally know if the Mata Menge had definite and indisputable evidence for early human colonization of Flores.

We decided to start work at Mata Menge, and hired a bemo from Boawae to take us to the geothermal hot springs, known as Air Panas (literally, "hot water"), near the village of Menge Ruda. From there it was an hour's walk across open country to the site, which is located on the west side of a hill alongside a track used by local herders on horseback passing through with herds of cattle and buffalo. Despite the remoteness of the area, people still know what is going on in the basin, especially when outsiders are involved. Soon after we arrived at Mata Menge, as we were starting to clear the vegetation that obscured the layers of sandstone and tuff, we spotted a figure on horseback about two hills distance and coming our way. In the middle of nowhere, Kornelius Podha, a local

Cultural Heritage officer from Menge Ruda, had come to check our documents to make sure that we had all the proper permits from Kupang and Bajawa. Kornelius became very interested in our work and ended up joining the team for the next three days to help draw up cross sections and take samples. His horse also proved very useful as a pack animal for transporting finds back to our rental vehicle. Although small, such "Flores ponies" have the spirit and appearance of diminutive Arab horses, and are enormously strong for their size.

You could still see the large area excavated by Verhoeven in 1963 and the smaller-scale work Aziz and his colleagues carried out in 1994. At the site we found the situation exactly as Verhoeven described it; layers of sandstone and lenses of white tuff about 1.3 meters thick contained fossilized remains of large *Stegodon*, crocodile and giant rat, as well as stone artifacts. Beneath those layers were pinkish tuffs that had been deposited as lake muds soon after the eruptions that produced them. The pink tuffs were largely devoid of fossils and stone artifacts, but were excellent dating propositions because the minerals in them were largely the result of single volcanic events. In contrast, the sandstones comprised materials from different sources and ages and so were poor dating prospects.

We determined the ages of the Mata Menge tuffs with the fission track (FT) method that analyzes single crystals of zircon within the volcanic tuffs for microscopic evidence of tracks caused by the fission of uranium atoms. When the uranium atom splits, the two fragments recoil from each other and cause damage in the structure of the zircon crystal, leaving tracks. Once etched with acid these tracks are visible under an optical microscope. Only extreme heat, such as that found in the molten magma of a volcano, can erase the tracks, providing a clean slate. After the rock has cooled, the number of tracks in a crystal grows with time, so counting the tracks and measuring background radioactivity can be used in combination to estimate age.

We cut out two 30-by-30-centimeter blocks of Mata Menge tuff for FT dating. Paul O'Sullivan, an expert in the technique then at La Trobe University in Melbourne, found that the pink tuff underlying the deposits of fossils and artifacts was 880,000 years old, while an overlying white tuff was about 800,000 years in age. Verhoeven had also been right about the approximate age of Mata Menge! People did make it to Flores by at least 800,000 years ago and possibly even earlier, long before most other researchers thought it humanly possible.

As if to emphasize the point, during the work we found a flaked stone artifact on the interface between the sandstones and the older tuff. We carefully removed it in a block of sediment. The ultimate bottom line for other researchers in accepting controversial new findings is the same for all of us—"seeing is believing"—and having an indisputable artifact in preserved stratigraphic context as a show-and-tell item would prove to be very useful.

After spending a few days at Mata Menge, we moved on to the older site of Tangi Talo, where fossil remains of pygmy *Stegodon*, giant tortoise, Komodo dragon and giant rat had been found in a 30-centimeter-thick layer of volcanic ash. It seems that a large number of animals had gathered around a water hole during a volcanic eruption, died there and been entombed in ash—another good dating opportunity using the fission track method. We later obtained an age for the tuff of 900,000 years, which tallied exactly with age that the Indonesian-Dutch team had obtained with the paleomagnetic method.

On the face of it, our results also indicated when humans first arrived in the region: stone artifacts were not present at Tangi Talo, but occurred in abundance at Mata Menge. This suggested

Endemic fauna on Flores included pygmy Stegodon, *giant tortoise, Komodo dragon and giant rats. These animals, evident at the fossil site of Tangi Talo in the Soa Basin, well illustrate the depauperate, unbalanced nature of fauna on this isolated island, and the size changes that often occur. Around 900,000 years ago, pygmy* Stegodon *and giant tortoise disappeared, and were replaced by large-bodied* Stegodon *around the same time as the first definite evidence for hominids—stone artifacts, as abundantly evident at sites such as Mata Menge.* (CREDIT: MIKE MORWOOD, AFTER MORWOOD, 1998)

that humans arrived on Flores sometime after 900,000 years ago, but before 800,000 years ago, a time that seemed to coincide with the extinction of pygmy *Stegodon* and giant tortoise, which, by Mata Menge time, had been replaced by large *Stegodon*.

Getting credible radiometric dates from these two sites provided the makings of a paper for the science journal *Nature*, in which we argued that humans, presumably *Homo erectus*, had reached Flores sometime between 880,000 and 800,000 years ago. In turn, that paper provided the basis for a successful grant application to the Australian Research Council (ARC) to begin larger-scale investigations in the Soa Basin, which ran from 1998 to 2001 and set the scene for later projects. Our test excavations at Mata Menge and

Tangi Talo were very small, but were strategic and produced highly leveraged results—small investment and large return. There were lessons here for future research.

For the following field seasons in the Soa Basin, we mapped the geology of the entire basin; excavated at the fossil sites of Dozu Dhalu, Boa Lesa, Kopowatu and Tangi Talo; surveyed the basin to locate more than 30 fossil sites; dated major geological layers; and pushed back the age of hominid occupation of Flores to more than 840,000 years ago. Meanwhile, the number of people and disciplines involved increased. Doug Hobbs was keen to continue our long-term research collaboration, and fitted in well with Aziz and his GRDC colleagues; and because the excavated finds included stone artifacts, as well as animal fossils, two archaeologists from ARKENAS, Jatmiko and Nasruddin, were invited to participate.

We made good friends with some of the Ngadha who worked with us, and also with some of the Nagakeo, a related ethnic group based around Boawae. Our main contacts in Boawae were Musa Bali, a distinguished and well-spoken man in his 60s, and his two sons Flori and Ferri, who seemed to be involved in all sorts of entrepreneurial, and often shady, deals involving rice, transport, antiquities and anything else they could make money from.

We were also helped by three brothers from Wolo Wavu, a small village close to Boawae, but down a very rough track that discouraged outside visitors. All three brothers, Alex Gadhu, Mingus Siga, and Ginus Denga, were very Melanesian in appearance. They were middle-aged and thin, with dark skin and fuzzy shocks of hair. But there the resemblance ended. Alex, the village headman, was a quiet and responsible man who cared deeply for his

Ginus excavating Stegodon *remains at Dozu Dhalu in the Soa Basin.* (PHOTO: MIKE MORWOOD)

aged mother. He organized the number of workers from his village we needed each day for the excavations. Minggus was much more vocal and emotional, sounded like he was angry even when he wasn't, delighted in finding fossils, and with his missing front teeth, head scarf, sarong and parang, he could have passed easily for a pirate. Given any opportunity, Minggus would search the hills and gorges of the basin looking for fossils eroding out. Their older brother, Ginus, a taller man, was the local expert in making palm wine. He was somber, and with bloodshot eyes he often seemed to have overindulged in his product. Ginus would squat all day in the blazing sun, methodically smashing up lumps of rock with a hammer, then sorting through the debris to extract artifacts and fossils ranging in size from bits and pieces of *Stegodon* right down to tiny rat bones and Komodo dragon teeth.

The families of the rajah, Musa Bali and the brothers at Wolo Wavu were all related in ways we never really understood and, away from our shared work, site recording, surveys and excavations, we only caught glimpses of Nagakeo life. Alex once casually pulled out of his pouch a small ground stone with very unusual wear: he told us it was used for filing the front teeth of girls during their puberty ceremonies. And just off the track into Wolo Wavu, we saw small stone slabs on which chickens were sacrificed prior to hunting expeditions, and the tree next to the altar was festooned with the jawbones of pigs. Every year in July there is also the first day of the hunting season, during which the countryside swarms with hunters on horseback pursuing pigs and deer, accompanied by dogs and armed to the teeth with spears, harpoons and parangs.

During one fieldwork season, the Soa Basin research team was invited to the ceremony that takes place on the evening of this first day of hunting. Since pig meat was to figure on the menu, our Muslim colleagues could not participate, so only Doug Hobbs and I went. The ceremony was held in a grassy area surrounding a large fig tree, against which participants had propped hundreds of spears and harpoons. Beneath the tree there was a fire tended by the master of ceremonies, who danced and chanted wearing a headdress decorated with pig tusks. Around the edge of the clearing about a hundred men sat, while a similar number of women sat in a group behind. We took our places with the other seated men, while a communal meal of rice and pig meat was served up from large wooden tubs onto wooden plates for each participant, and then washed down with cups of palm wine. Pig meat is considered a real delicacy for the Nagakeo, as it is for most other ethnic groups in Flores, but the wild pig meat that was served up that night had the flavor and texture of patched truck tires. It was impossible to chew, let alone swallow. The best we could do was suck on lumps of the meat until we figured no one was looking and surreptitiously toss them over our shoulders into the darkness beyond. We did not see until too late what was plainly obvious to others, that a pack of dogs had positioned themselves behind us specifically to take advantage of these tossed tidbits.

Work in the Soa Basin was hard. The routine for six days a week was for us to leave the house of the rajah around 5 a.m. in order to be able to drive for an hour, and then walk for another hour to the sites in the cool of the early morning. During our excavations on the east side of the Soa Basin, at Dozu Dhalu and Kopowatu, the village of Wolo Wavu was the starting point for the walk up and over a steep ridge to the sites. We had to allow at least two liters of water a day per person and, because of the absence of shade, had to erect tarpaulins for some protection from the blazing heat. The deposits at these sites are rock strata of varying hardness and our excavation tools included sledgehammers, large chisels, picks, screwdrivers and trowels. Large slabs of rock were detached in 10-centimeter levels called spits, then systematically smashed up into smaller fragments by our workers. We tried sieving, but by the time the rock matrix had been smashed up small enough to pass through a three-millimeter mesh, any bones or artifacts had been similarly pulverized.

We employed both Nagakeo and Ngadha workers, who took the physical demands of fieldwork in the Soa Basin in their stride.

They could carry impossible loads long distances, climb steep slopes and spend hours smashing rock with heavy hammers and picks, all with good humor. And if we encountered a wild pig during our travels, everyone would down tools, whistle up the dogs, pick up spears and charge off in pursuit. Minggus had a particularly good hunting dog, Hero, which accompanied him everywhere. Unfortunately, in 1998, an outbreak of rabies on Flores killed at least 113 people, and the local authorities ordered the mass culling of dogs, including Hero. Minggus wept openly.

At the end of the day around 4 p.m., we would stagger back to Wolo Wavu, then spread out on bamboo platforms under shady trees to drink green coconut milk and sweet, spiced coffee. Our general observation in Flores was the poorer the village, the better the coffee, and the Wolo Wavu coffee is very good.

Digging at Mata Menge, Boa Lesa and Kobatuwa on the west side of the basin meant that we could walk back to Air Panas, arrive exhausted, then soak in the world-class hot springs. Steaming hot water bubbles out at one end of a pool about 20 meters across, then down a roaring creek and into a plunge pool, where another cold-water creek joins it. You can pick an appropriate place to bathe depending on whether you want a hot, warm or cold soak—or move between. Stripped to the waist, I would ease myself into the pool, find a suitable rock shelf to lean back on, relax, and let the stresses and strains of the day wash away.

We were interested in all periods of human occupation of the region, from initial colonization until the present day. So Tular Sudarmadi, a lecturer from the University of Gadjah Mada in Indonesia, who was doing a postgraduate degree at the University of New England in Australia, joined our research team. Tular carried out an ethnoarchaeological study of the Ngadha, with particular emphasis on the social, ideological and economic role of their megaliths. As part of his research he surveyed and mapped four Ngadha villages, including all houses, walled plazas, megaliths, other religious structures, granaries, graves, toilets, pig sties and garbage disposal areas.

Tular also planned to use his information on the distribution of activities and structures in modern Ngadha villages to guide the excavation of an abandoned Ngadha village, Wogo Lama. This would be the first archaeological investigation of the Ngadha, who have stories about how their ancestors came to Flores and where they settled. We had a meeting with Ngadha elders at Wogo Baru to discuss excavation options and gauge their attitude to such research. Provided it was done with respect, community involvement and feedback, and with the proper sacrificial ceremonies beforehand, they were very supportive. In fact, they looked forward to the start of research on their history. Most of the men who spoke that evening were elderly, but there was one younger man, Siprianus Batesoro, dressed in traditional garb like the others. He translated the Bahasa Indonesia discussions into Bahasa Ngadha for the benefit of the elders, but otherwise said little until we circulated some of our publications on the Soa Basin for people to look at the photos. He then asked in perfect English about the type of humans likely to have been in the Soa Basin 840,000 years ago. A little surprised, I asked Siprianus where he had learned English. "At the University of Gadjah Mada during my doctoral research on nuclear physics," was his reply. Archaeology is full of surprises. Siprianus later became United Nations coordinator of the resettlement program for East Timorese refugees, as well as a friend, and he is still interested in our research.

The major problem with research in the Soa Basin is that the deposits containing fossils and artifacts stop around 650,000 years ago. The record does not pick up again until very recent times—historic Ngadha sites directly overlay deposits of much greater age. It was like whole chapters had been ripped out of a book at a crucial moment, the story only reappearing near the end, long after the climax. To find out what happened in the intervening period, we began to visit other parts of Flores with different geological deposits of different ages. While in the Soa Basin, I learned that Jatmiko had previously excavated on Flores with Soejono at a large limestone cave called Liang Bua, one of many sites excavated by Father Verhoeven during his 17 years as a resident priest on the island. So in 1999, while our work at Boa Lesa, Tangi Talo and Kopowatu in the Soa Basin was still ongoing, a small group, including Aziz, Doug, Jatmiko and myself, drove to Ruteng, the Manggarai provincial capital in western Flores, to visit Liang Bua. Jatmiko acted as guide.

Stepping into the cave for the first time, I was immediately struck by its size, and particularly impressed by its suitability for human occupation: it was spacious, well lit with a northern outlook, and had a flat, dry clay floor, which would have made it a comfortable place to live. We also visited Liang Galan, another nearby cave, which is similar in overall size and morphology to Liang Bua, but has a sinkhole at the base of the main chamber and has not retained deposits. The empty chamber at Liang Galan is about 17 meters deep, and it seemed likely that Liang Bua could contain a similar depth of clay deposits. Soejono later told me that he had obtained a radiocarbon age of 10,000 years from a depth of three meters. The mathematics was compelling. There was every chance that the archaeological sequence could provide evidence for when modern humans first arrived and what had happened to the descendants of the people who made the Mata Menge stone tools so long before. It was a very exciting excavation prospect—the best I had seen in more than 30 years as a professional archaeologist.

Returning to Jakarta from the 1999 Soa Basin fieldwork, I initiated discussions with Soejono about carrying out further excavations at Liang Bua to plumb the depths of the site, and to analyze, date and publish the huge backlog of material that had accumulated from previous work there. Initially cautious, he gradually warmed to the idea. The real breakthrough in our discussions occurred when we attended the Australian Rock Art Research Association Conference together in Alice Springs in 2000. Despite our differences in age, experience and perspective, we got on well and resolved to start work at Liang Bua the following year.

At the beginning of 2001, the last year of the Soa Basin ARC grant, I took six months' study leave from the University of New England to participate in two fieldwork projects. The first was to finish the geological survey and mapping of the Soa Basin with Aziz and his colleagues; the second was to commence excavations at Liang Bua with Soejono. The Soa Basin work ran for two months. As in previous years, I had applied for a research visa from the Indonesian Academy of Science (LIPI) to continue this established research program with Indonesian colleagues, but it was turned down ostensibly "because of security concerns," but probably because relations between Indonesia and Australia were then at an all-time low over unrest in East Timor. However, as an Indonesian research institution, GRDC did not need LIPI permission. Aziz and his fellow researchers were therefore able to carry on and complete the final season of fieldwork in the Soa Basin. Fortunately, I was able to briefly visit the Soa Basin in March 2001 to discuss fieldwork results with my GRDC colleagues.

Back in Jakarta, Soejono and I began preparations for excavations at Liang Bua—to be undertaken under the authority of ARKENAS, another Indonesian research institution. The director, Dr. Haris Sukendar, provided a formal letter inviting me to participate in the work, while Soejono arranged for three experienced fieldworkers—Thomas Sutikna, Rokus Awe Due and Sri Wasisto—to join the expedition. They had worked with him many times before, and were enthusiastic about the new prospect. Soejono also contacted the Kupang and Manggarai administrations, as well as friends in Ruteng, to ensure that people knew we were coming, and that there would be no problems in getting local permits or people to assist with the work.

We discussed excavation options for the site and decided that Sector IV, an area in the center of the cave's main chamber, was the best prospect because of the high density of artifacts and bones that had been recovered there previously in the uppermost levels. Soejono had excavated this sector to a depth of around four meters and thought that deeper deposits would probably not contain artifacts, but, to his credit, he was willing to be proved wrong. The planned work would be a difficult operation involving the removal of 33 tons of backfill from the previous excavations, and timber shoring of the trench walls, which had slumped when previously left open for a year. If this proved too difficult, the fallback plan was to redig another area against the east wall of the cave, Sector VII, where Soejono had found a number of human burials.

On April 10, 2001 we flew to Kupang in West Timor to get permits from the departments of Culture, Police and Social Politics for the planned excavation at Liang Bua. At Kupang we were picked up at the airport by local staff from the Department of Culture in official cars and chauffeured around various government offices. Soejono is regarded as the revered elder statesman for archaeology in Indonesia and was treated accordingly—not surprising since he had trained almost every other Indonesian archaeologist of note. He had also worked with Professor John Mulvaney in Sulawesi more than 30 years ago. John had been my Ph.D. supervisor at the Australian National University, so there was a chain of connection here. After spending several days in Kupang, everything was in place and we set off to Flores to start our first season of work together at Liang Bua.

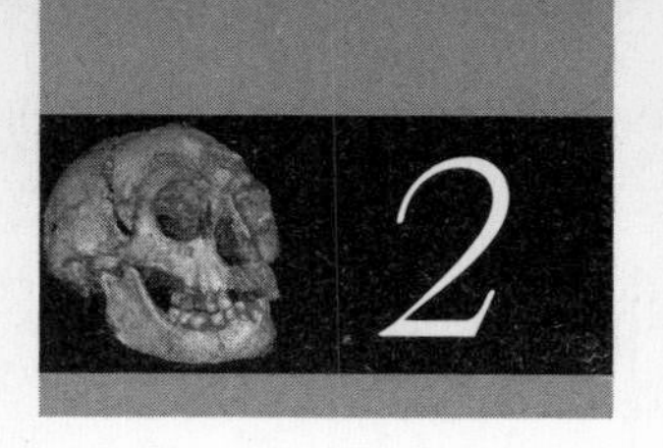

Stories from Cold Cave

Liang Bua is a very easy site to work on. No long treks with heavy loads required. In Ruteng, just get on one of the trucks with plank seating, which serve as local transport for people, animals and goods, and after an hour on the rough but scenic road down to the village of Terus, you clamber off right at the cave entrance. The cave is about 50 meters wide and 20 meters high, but partially hidden by coffee trees. It is not until you step inside that you realize how big it is. An intrusive concrete archway and path now installed by the Manggarai provincial government leads up to a lockable gate in the high mesh and barbed wire fence that restricts access to the cave. The key is kept by the official cave guardians Rikus Bandar and his son Agus Mangga, who also act as guides for the few tourists venturing this far out of Ruteng, the provincial capital.

Inside, the cave is cathedral-like, a flat clay floor with great chandeliers and cones of stalactites suspended above. They are huge and many are impossibly bent toward the light at the cave entrance. How some of them remain up is a mystery. Occasionally they don't. Fallen stalactites litter parts of the cave floor, probably brought down in earth tremors so common to this region, one of the most volcanically active on earth. Even a small stalactite dropping 20 meters or so could kill, but we were always comforted by the fact that it doesn't happen all that often and that if one did fall and hit someone, they would never know about it.

The deceptively serene-looking Wae Racang River, 30 meters below and 200 meters distant from Liang Bua, is the force behind this landscape, slowly wandering backward and forward across the picturesque valley, cutting and carving as it goes, patiently wearing it down. It hasn't always been like this though; at times a sudden uplift of the area, or a drop in sea level, led to periods of rapid down cut, leaving remnants of alluvial plain perched high and dry. At least three such river terraces can be seen on the slopes down to the river. On the river flats, comprising the lowest terrace, there are verdant rice paddy fields, ingeniously partitioned by low, earth banks so that each field can be flooded when required. This area is the "rice bowl" of Flores where farmers get up to three crops a year. Higher up, coffee and bananas are grown along the second river terrace, intermingled with plantings of yam and sweet potato. Higher up still, the rugged, limestone hillsides towering over the valley are increasingly being cleared and terraced for dry field cultivation, particularly of corn.

The downcutting of the river also left behind a stacked series of caves connected by sinkholes; above the present-day Liang Bua, for instance, there are the collapsed remains of an earlier cave, and below at the water table is another cave, still being formed. Similarly, although it is now spacious and airy, Liang Bua was in fact conceived underground at least 380,000 years ago in a sealed womb of flowstones, stalactites and stalagmites, all formed by the slow drip of calcium-laden water. Around 200,000 years ago, the river finally breached the cave's walls with such an explosive inrush of water that it carried rocks, cobbles and even large boulders into the cave. Cemented into this jumble of rocks, or conglomerate, were stone artifacts, evidence that people were already living in the vicinity at the time. The river drained away and, slowly but surely, water falling from cracks in the ceiling began to erode the conglomerate. Boulder by boulder, water pried underneath the thick layers of perpetually accumulating flowstone causing them periodically to collapse to form the great sloping slabs along the west wall and rear of the cave on which we eat, rest and smoke the local clove cigarettes during breaks.

A massive volcanic eruption about 100,000 years ago filled the Wae Racang River with loads of silt that were dumped in the cave in layers up to eight meters thick. These layers show evidence of cutting, infilling and distortion over a long period. The earliest evidence for humans using the cave so far dates to around 95,000 years ago and comes from an overlying layer of reddish brown clay, largely derived from weathering of limestone. Gentle slope wash from crevices at the rear of the cave is still depositing these clays that in some parts are at least 11 meters deep. About 12,000 years ago, another massive volcanic eruption produced a huge cloud of ash that blanketed the area around Liang Bua, with catastrophic effects on local vegetation and animals. This eruption sent another surge of distinctive white silt into the cave, like a shroud covering evidence of its former occupants and their lives, all buried or cemented in place; trapped in the cave's own inexorable evolution.

The people around Liang Bua, the Manggarai, are of varied physical appearance. Some could be from Bali, others from the Highlands of New Guinea, and their own history confirms that they are of mixed origin. Ancestors of the royal Todo clan, for instance, only migrated to Flores in the 17th century, and the name Manggarai (which literally means "the anchor is running") refers to a magical anchor on their vessel that would drop on its own accord to indicate where people should disembark. It is also recorded that the long-established aboriginal population resented the immigrants' acquisition of land and growing power, leading to a 100-year conflict before final victory of the Todo alliance in 1750 CE centralized political control over the whole Manggarai area.

Liang Bua must always have been a special place, and for the last 10,000 years the dead have been placed here with ceremony. From that time, local inhabitants have carried out multistaged rituals for disposal of the dead. The bones of the deceased were collected and bundled together, smeared with ochre, adorned with ornaments, such as marine shell necklaces, and cached in the cave. During the Neolithic period, when cereal cultivation, pottery and ground adzes were introduced to Flores about 4,000 years ago, one of the main functions of Liang Bua appears to have been for burial of the dead with grave goods, including pots, tools and ornaments. Later, after the first appearance of metal at the site around 2,000 years ago, mourners sometimes placed highly valued iron and bronze tools next to the body of the deceased. And traditional beliefs about ancestral spirits are still apparent today in the offerings of eggs and food placed at the rear of the cave by persons unknown.

Beliefs about the beginning of the world and how life originated determine many aspects of Manggarai life, including their traditional ceremonies, religious structures and the layout of fields, villages and houses. Like most other people on Flores, the Manggarai are now overwhelmingly Catholic, but blend Christian beliefs with traditional *adat* beliefs that place emphasis on ancestral spirits and offerings. The Manggarai say that the first people emerged out of the strong, versatile bamboo plants growing on a newly formed Earth, which, after a great collision, was lifted streaming from the sea that covered the entire world. Back then, Earth, the mother, was connected by a vine to Sky, the father. But mother and father flew apart when a dog bit the vine separating them. Human beings—who are said to have originally had long fur—spirits and animals were closely related then. Humans became distinguished from the other animals and forest spirits by their ability to cultivate plants. Finally, with the use of fire, came rules of eating and rules of marriage, and only then the transformation to human beings.

At first soft and malleable, the Earth hardened as it aged, trapping the impressions of ancestral activities and the sacred rituals associated with them. An ordered pattern of ritual was established, weaving together the connections between the spirit world and the human world, between clans, between human beings and ancestors, and between humans and the surrounding fields.

That is why the Manggarai still prepare special round, segmented fields (*lingko*), the centers of which are clearly marked with a ritual rock where chickens are sacrificed to the spirit living at the spot. A central altar, representing the source of the seeds, is arranged. A branch of a special tree, representing the penis of Father Sky, is planted near the rock, and both are encircled by vines, representing the womb of Mother Earth. Such a fertile center runs the risk of becoming overheated, and water is placed under the rock to cool the garden. A piece of iron (thought to be what sustained the ancestors before the advent of food)—and perhaps even a dog—is also buried. The altar recalls the stories of the golden era when food was created by the ancestral father, "Adam," who cut up his first-born child into pieces to plant in the garden. He used the branch of a certain tree as the slaughtering knife, first placing the child on the rock. Seven days later, according to myth, rice and maize started to sprout, the shoots calling out to the understandably upset "Eve" that they were the child reborn in a new form. In fact, all traditional Manggarai ceremonies serve to remember this life-sustaining sacrifice of the first human child, and blood sacrifice is a pervasive part of Manggarai culture.

For instance, whip fighting (*caci*) is a distinctive, iconic Manggarai ritual performed at the end of the rice harvest. To the accompaniment of gongs and great social excitement, pairs of men in ceremonial regalia take turns trying to cut the other with a buffalo hide whip, while the other defends himself with a buffalo hide shield. The idea is that the spilling of human blood at appropriate times will bring good fortune and ensure fertility of the soil for the next crop. There is great status in this and men proudly sport scars from previous duels. Occasionally someone is killed, but the person responsible is not punished.

The beliefs of priests and archaeologists were easily slotted into such an all-embracing cosmic worldview. The Dutch sent Catholic missionaries to Flores in the early 1920s, and today their success can be measured by the fact that most of the Manggarai are baptized Roman Catholic. But the Manggarai above all else are syncretists, accepting names and stories from the Bible and taking them as their own, blending them and molding them to fit their own stories. Adam and Eve easily became the names of the first people, as the beginning of the world became the Garden of Eden. The new names merely embellish the old stories. Even God is Manggarai.

It was religion that brought Father Theodor Verhoeven to Flores and then to Liang Bua, which was in use as an elementary school when he arrived. Bamboo benches and tables were set up for the pupils and a teacher would supervise proceedings standing on the natural platform afforded by the conglomerate block at the rear of the cave, with his hands behind his back, looking down on his charges while lessons were recited. During a visit to

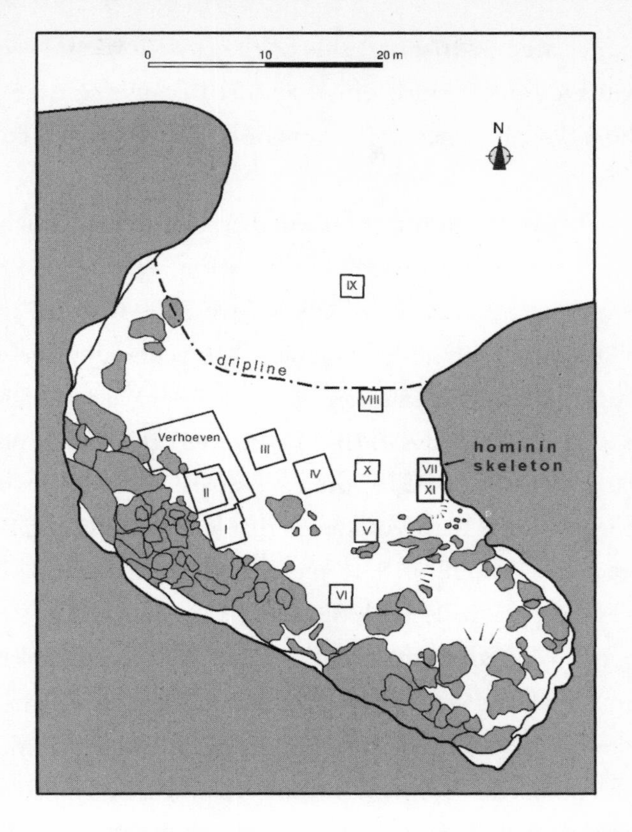

Plan of Liang Bua showing the locations of excavated areas. Between 1978 and 1989 Soejono excavated Sectors I to X, but only to a maximum depth of 4.25 meters. We continued the excavations in Sectors I, III, IV and VII, and began a new sector, XI, to reach a maximum depth of 11 meters. Bedrock has still not been encountered in any of the excavated areas. (CREDIT: MIKE MORWOOD)

Liang Bua, Father Verhoeven decided the floor of the classroom was a great excavation prospect, and in 1950, just after the school was moved to a more conventional classroom, he excavated a small test pit against the west wall just inside the entrance. The pit yielded encouraging amounts of stone artifacts and pottery. In 1965, over a two-week period and assisted by trainee priests from the Catholic seminary at Ledolero, Verhoeven excavated a much larger area on the west side of the cave floor and recovered six Neolithic burials with grave goods. Because of time limitations, however, his excavation was only taken down a couple of meters.

And so began the history of research at Liang Bua, which we are continuing, assisted by some of the same people who assisted Verhoeven—or by their sons. Rokus Awe Due, whom Verhoeven called a "diligent high school graduate," 40 years later does the initial classifying of bones and shells from our excavations. Sius Sambut and Rikus, schoolboys during Verhoeven's time, assist us now as Manggarai elders, and Agus, the cave guardian, is Rikus's son. It was correspondence with Verhoeven that led Raden Pandji Soejono, then director of ARKENAS, to begin work at the site in 1978, with further excavations in 1981, 1982, 1985, 1987 and 1989.

Soejono eventually excavated a total of 10 squares, or sectors, at Liang Bua, and uncovered a spectacular cultural sequence. In fact, it is the only excavated site in Indonesia so far to yield evidence for Paleolithic, Mesolithic, Neolithic and Metal Age occupation. This evidence included artifacts of stone, bone, shell and metal; pottery; burials with grave goods; and a wealth of well-preserved animal remains. Soejono excavated to a maximum depth of 425 centimeters, reaching some 10,000 years into the past—to the onset of present-day climatic conditions, an epoch termed the Holocene, when farming and the first stirrings of what we call "civilization" began. He did not reach bedrock, however, in any sector, so the total depth of deposits and the full history of human use of the site remained unknown. Nor were any of his excavation findings published.

When we began our first excavation at Liang Bua in mid-2001, we aimed to take one of the previously excavated sectors to bedrock. Although Soejono believed that the deepest levels of his ex-

Raden Pandji Soejono, Mike Morwood and Sri Wasisto at Liang Bua in 2001, our first field season of work together. (PHOTO: MIKE MORWOOD)

cavations were culturally sterile—that "the deposits had gone quiet"—there seemed to me every possibility that the archaeological record might continue much deeper and help fill the gap in Flores archaeology between that of the Soa Basin and the present day. The last 100,000 years would be of particular interest because the little evidence available from Southeast Asia indicates that the arrival of modern humans and extinction of earlier human populations is likely to have occurred during this period.

There is a formality in the Liang Bua excavation based on mutual respect that seems to work well. All adult males are referred to as Pak, all females as Bu—we are all performing important roles in the work and our senior workers are also elders in the Manggarai community. There is formality also in the way that the communities here welcome and farewell the research team at the beginning and end of the field season. Each village sends a delegation led by the village chief, accompanied by elders (both men and women) and younger people of standing—about 15 people in total. All are dressed in fine traditional attire and bear woven mats to sit on, cushions, a chicken to offer to ancestral spirits and a gourd container of palm wine. We pray for good fortune in the coming field season. Speeches of mutual goodwill are made, and cigarettes and palm wine are consumed before the meeting breaks up with social chitchat.

The cave is large, relatively dry and cool—hence its Manggarai name, "Liang Bua," or "Cool Cave." It is a pleasant place to live or work when it is hot or wet outside. There is enough space to excavate, sieve what we dig, process finds, store bamboo and timber for shoring, ladders and tables and set aside areas for resting and eating, all without having to step out into the tropical heat. When things are running well, there is a quiet, busy hum about the work—the distant clatter of the generator set up outside the cave provides electric lighting to the diggings; the sound of people who know their jobs quietly getting on with them; the movement of bucket carriers shuttling from the excavation to the weighing scales and sieves; conversations around the benches set up for eating; preliminary analysis of finds and writing of reports.

There is also a constant flow of visitors, local people on their way back from markets, dressed in the traditional colorful Manggarai sarongs, groups of curious children so numerous in this area that seems to be undergoing rapid population growth, and the occasional official from the Manggarai administration. This is not an inconvenience, but welcomed as an opportunity to explain what we are doing and why. It's important that the local people know we are actually digging up bones and stone tools, not gold or other treasures! There are many precedents where archaeological work has gone awry. For instance, on one occasion during his excavations at Liang Bua in 1965, Verhoeven had to accompany police back to Ruteng for questioning. Local officials thought that he was digging without proper permits—and he ended up spending a night in a cell of the old Ruteng prison.

The excavators dig methodically, using bamboo stakes, trowels, metal-edged digging sticks, picks or crowbars, depending on the circumstances—at least six hours each working day squatting on their haunches doing work requiring concentration and stamina. Not many of my Australian students could do this, but as rice farmers the Manggarai are used to hard physical work—and some, such as Benyamin Tarus and Ansel Mus Ganda, developed real expertise in excavating by stratigraphic layers while plotting the exact position of finds.

At any one time, there may be four people working in a two-by-two-meter square. Digging by 10-centimeter-depth units or by layers, whichever is the smaller, they also plot and bag finds—stone artifacts, bones and charcoal concentrations. Buckets are hauled up to the surface using long bamboo poles with a side shoot left as a hook. Bamboo is a remarkable material, incredibly versatile, strong and cheap. People working in the excavation must keep watch when a bucket is being raised—a handle could give way. They all wear industrial helmets—even a small rock or tape measure dropped down an 11-meter excavation shaft could seriously injure or even kill.

On the surface each bucket is first weighed, then dry sieved, and the material retained in the sieves is carefully sorted by people with nimble fingers and sharp eyes to recover more artifacts and faunal remains. However, adhering clay makes it difficult to identify many artifacts and bones, so what's left in the sieves is then rebucketed and carried on the heads of other workers out of the cave, down the track and out into a paddy field, where a wet-sieving operation has been established. Ingeniously designed by the Manggarai, who are masters at moving water around to irrigate fields using small dams and networks of bamboo pipes, water cascades out of a bamboo pipe into a small sump pit. Our two local wet-sieving specialists, Gaba Gaur and Rius Laru, sit there with a sieve into which buckets of excavated clay are deposited. The water washes the artifacts and bones clean, and they are then picked out, dried, bagged and labeled. This increased the number of finds from the excavations tenfold.

Initially wet sieving was undertaken in the river—not good for the river or those using it downstream and this sieving strategy also resulted in loss of the clay, which we needed to make up when backfilling. In the new system, clay passed through the sieve and collected in the sump, which was cleaned out regularly and the washed clay allowed to dry. The stream of workers bringing buckets from Liang Bua for wet sieving then took back buckets of sieved, dried clay, which was used to backfill the excavations at the end of the field season.

In fact, processing up to 200 tons of deposit in each season at Liang Bua required a production line system that took into account different rates of progress in the excavation and in the processing of excavated materials. When the excavation progressed rapidly, piles of dry-sieved deposit, each with its own label and corresponding to an excavation unit, were laid out on plastic tarps on the cave floor. When the excavations slowed down, the wet-sievers caught up and the backlog of piles diminished.

With all artifacts and bones removed, some processed deposits were put into rice sacks and used to make a walled enclosure into which the bulk of deposits were placed. When that enclosure was full, another enclosure of rice sacks filled with deposit was started on top of the resulting platform, and so on. This kept the workspace within the cave relatively tidy, despite the sheer bulk of excavated clay. By the end of the digging, the tiered heap of spoil near the mouth of the cave looked suspiciously like a stepped temple, which took 20 Manggarai workers three days to demolish when backfilling the trenches.

The hard physical work of the excavation depends on Manggarai muscle power. The wages paid are low by our standards, but still double what local road workers earn. If we paid any more, we would set up a salary benchmark for the area that other employers could not match and there would be problems. Also, overgenerosity could be taken as a sign of stupidity or great wealth. Either perception is to invite trouble—a fact that applies anywhere, not just in Flores. One of the few problems we have had is that too many people wanted to work with us on the excavation. We employ about 40 people but could easily recruit several hundred more if required. Locally, there are not too many other options for earning the cash required to buy agricultural tools, medicines or children's schoolbooks—especially during the dry season, when little cultivation is done. We also provide cigarettes, bananas, yams, corn, coffee and moderate amounts of palm wine, which is an important part of the diet, for the morning and afternoon breaks. In fact, at times the excavations seem to run on these commodities.

With wages, purchase of bamboo, planks, foods and equipment, accommodation costs and hire of transport, our research actually makes a major contribution to the local economy, which is one of the reasons that people are so supportive.

Beyond the economic, our research makes other contributions to local life, too. As with the rest of us, the Manggarai enjoy working and socializing with people from very different backgrounds, and learning new things—especially when it involves digging deep holes to uncover stone artifacts, the remains of animals like *Stegodon* and the occasional human. They take great pride in their work and findings. When nearby Terus village fielded a soccer team to play against another village, the players' jerseys were embellished with a *Stegodon* motif.

Work starts at eight in the morning and finishes at four in the afternoon with an hour off for lunch and coffee breaks at ten and three. At lunchtime, the workers break up into groups according to village or family around the edge of the cave. Wives, accompanied by children, bring in lunches for their men and remain to chat. Some workers have a nap on plastic tarps, sloping slabs of rockfall or in berthlike small side caves. Indonesian and Australian researchers tend to congregate around the wooden tables and benches set up near the rear—eating packaged lunches brought from a Padang restaurant in Ruteng, discussing the day's activities and finds.

Ibu Grosa, the wife of Rikus, is employed to make the coffee and provide snacks for the breaks. She is a middle-aged woman, younger than Rikus, maybe 35, and could almost be Polynesian with her strongly chiseled features, high cheekbones and long, straight hair tied back. She comes to the site with a large kettle of strong, sweet Flores coffee, and woven bags containing enamel cups, and maybe boiled yams, or sections of roasted corn on the cob. If Grosa arrives early, she squats just inside the cave entrance and waits patiently. It's hard to know what she thinks of the proceedings. Rikus, as a Manggarai elder and site foreman, stops his recording of bucket weights and announces each break with a loud clapping of the hands. He is in his sixties and no longer in good health, but he is certainly a man of influence and dignity. In contrast to his wife, Rikus is thin and wiry, with dark skin and tight curly hair—overall very Papuan in appearance.

With the break called, dry-sievers and sorters working on the surface within the cave stop and quickly gather around Grosa for coffee, snacks and the rough clove cigarettes that most favor. Those down the excavation pits have to make their way up a series of bamboo ladders and platforms to the surface. With their hard hats, sweat and soiled clothing, they look like workers on a construction site, and in many ways that is what they are. Bucket carriers also straggle into the cave from the wet-sieving area. The breaks serve not only for rest and refreshment, but as social gatherings of villagers during which much animated conversation takes place under a canopy of dense, clove-scented tobacco smoke.

Sometimes during the quiet of a break, Agus will start singing a traditional Manggarai ballad of love unrequited or lost. His voice has a soprano-like, ethereal quality that chills the spine. It fills the cave and people stop whatever they are doing to listen. We visitors don't know the exact meaning of the words, but the general sense is clear enough. Early Catholic missionaries reported that wherever they went in Flores there was the sound of singing and of flutes. Now youngsters in urban centers such as Ruteng, Bajawa and Maumere mainly listen to Western-style pop.

The Manggarai are polite and sociable, but there are limits. We first learned this with Pak David, a retired schoolteacher who occasionally helped out on-site. Apparently because of his education and position, he regarded himself as better than most of his fellow workers and showed it. Unknown to us, this was greatly resented. One morning as we commuted to Liang Bua, one of our workers, Pit Ludu, gestured from the side of the road and indicated that our truck should stop. There was a hurried exchange and David, whom we had picked up en route, got off the truck and hurriedly began walking back home. We were told that if he had gone to the site on that day there would have been serious trouble.

Just how serious the trouble could have been became evident some weeks later in Ruteng, where I'd gone to the Wartel, the local telephone exchange, one night to phone Australia. The lobby of the exchange was full of people, some waiting for a telephone booth to become free, others just socializing. While making the call, I suddenly heard loud whacking sounds and people running out of the exchange. Three Manggarai villagers, armed with parangs and spears, were attacking a well-built man at the other end of the lobby. A parang slash to the side of his head and he stumbled forward and sprawled just outside my booth door. The attack continued about one meter away. If they started on me, could I deal with them? The three assailants ran outside, but came back to administer more kicks, spear thrusts and slashes. One bent down and with his parang carefully sawed through the victim's ankle sinews. They then ran out again, this time for good. I was stunned and had to step over the unconscious victim and streams of blood to get out of the booth. He moaned softly. There was no other sound. One of the Wartel staff poked his head around the corner. "Get the police," I said. "He is police" was the reply. A minute later uniformed police ran up from their nearby compound. They were seriously angry and armed with assault rifles. It was definitely time to leave.

The attack in the Wartel was a local matter involving a woman, but disputes over land ownership are probably the main cause of violence. In addition, relations between villagers and police are not good. The year before, in 2002, police had begun evicting people and uprooting plantations of coffee trees on traditional village land that had been incorporated into a state-owned protection forest. Matters came to a head in early 2004, when about 500 Manggarai farmers armed with spears, parangs and rocks protested outside the Ruteng Police barracks, and the police opened fire, killing six and wounding 28.

To commence work at Liang Bua in 2001, our team of five researchers and 12 Manggarai workers spent three hard days removing the backfill, bucket by bucket, from Soejono's previous excavation in Sector IV. With this done, we next constructed timber shoring—an essential safety measure as the excavation deepened. This was new to the Indonesian researchers, but they soon developed great expertise in designing and constructing shoring. The system is based on vertical planks supported by a framework of posts and cross-braced beams, with platforms constructed at two-meter intervals for safety and ease of access, and connected by bamboo ladders. All were made on site by Dius Nggaa, a local carpenter, who can knock up a table, seat, ladder or house on request using local materials and basic tools. Stratigraphic sections are drawn up and the shoring extended down as the work proceeds. When we are working deep, the excavation squares look like mine shafts, which is basically what they are.

One advantage of the system is that planks can be raised to view different parts of the excavation wall if need arises to check the stratigraphy or take samples. At the end of the field season, the shoring is then dismantled from the bottom up. If we left it in place, there is every chance that the backfilled soil would be emptied out by local people to recover the wood, which is recycled as material for houses and furniture. In fact, at the end of the excavation, our Manggarai workers carefully straightened every used nail, cleaned and neatly pressed each rice sack once emptied, and salvaged each piece of plastic tarp for reuse.

I first learned how to shore from two fellow Australian researchers, Rhys Jones and Bert Roberts, while excavating the site of Mushroom Rock in Southeast Cape York Peninsula with them in 1994. In turn, they had learned it by attending a grave-digging course in Sydney. Occasionally, graves are still excavated by hand and anything deeper than 1.5 meters must, under the Excavation Code of Practice in Australia, be properly shored. It is a fact that, with a few exceptions, previous excavations in deeply stratified limestone caves in Southeast Asia have stopped at around three meters depth or less—Leang Burung 2 in Sulawesi, Tabon Cave in Palawan, Niah Cave in Sarawak, Tham Lod in northwest Thailand and Lang Rongrien in southern Thailand, to name but a few. Why? Because digging any deeper is very dangerous and almost certainly illegal without shoring. There is, therefore, strong incentive for archaeologists to stop at "culturally sterile" levels when artifact concentrations start to tail off rather than continue excavations to bedrock. One of the exceptions, Song Braholo in East Java, was excavated to a depth of seven meters without shoring, but they were lucky and got away with it. Given recent findings, such as the modern human tooth excavated from the 125,000-year-old Punung III breccia deposit in East Java, it seems likely that the much-quoted date for the arrival of modern humans in Southeast Asia—some 50,000 years ago—is largely an artifact of limitations in excavation strategy.

Work proceeded haltingly at first, while we all adapted to new excavation and processing methods, and then gathered pace. We discovered that the previous excavations by Soejono had stopped just short of major changes in stratigraphy, stone artifacts and animal remains. In the 1980s, Soejono thought that he had reached the bottom of the cultural profile. But we kept digging and hit a thick deposit of white tuffaceous silt from a major volcanic eruption that had somehow been washed into the cave in large quantities. Sure enough, the ash was sterile, silent. But then immediately below the silt the deposits became noisy again—very noisy. Here were high concentrations of the bones and teeth of *Stegodon* associated with masses of distinctive stone artifacts. This was the first time that *Stegodon* remains had been found inside a cave in Indonesia, and eventually we were to recover skeletal parts of at least 47 individuals. The adults were about 500 kilograms (kg) in weight—the size of a modern water buffalo. However, most of the excavated remains were of very young animals—neonates, in fact—with milk tusks about the size and shape of clove cigarettes, and unworn milk molars in abundance. In the same levels, there were also the remains of bat, rat, Komodo dragon, other lizards, birds, snakes, frogs and fish.

We continued to dig, clawing our way back through time. In the northern part of the square, Rokus found a layer of rockfall cemented together by flowstones. Limestone is really peculiar. It is easily dissolved by water containing carbonic acid, formed when carbon dioxide dissolves in water, but if redeposited as flowstone, stalagmites or stalactites, limestone is incredibly hard. Now was not the time for trowels and bamboo stakes but sledgehammers, chisels, crowbars and picks. Finally, we broke through the flowstones and reached clay again. To our utter amazement, the layers of clay were packed with stone artifacts, bones and teeth—up to about 5,000 artifacts per cubic meter of deposit. Everyone was

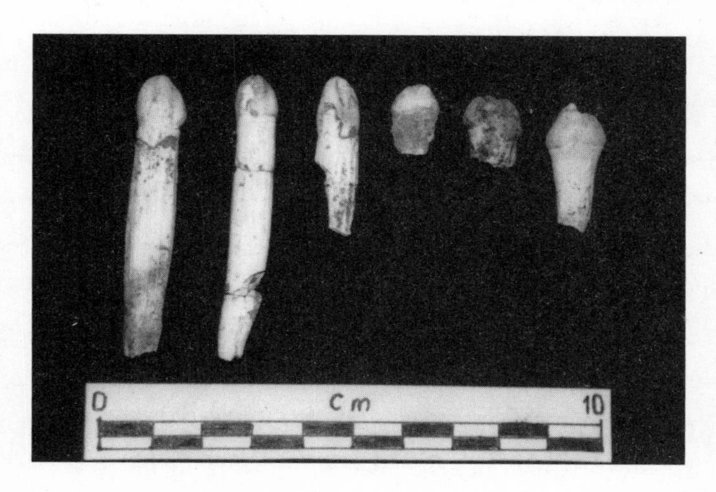

Stegodon *milk tusks and molars.* (PHOTO: MIKE MORWOOD)

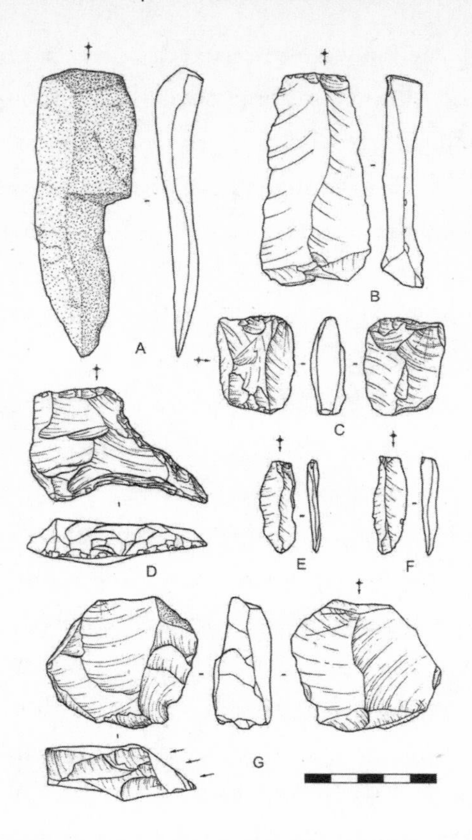

Stone artifacts found in association with Stegodon *remains at Liang Bua.* (CREDIT: MIKE MORWOOD)

rapt with excitement. Here was evidence of human activity by the bucket load. It was simply extraordinary. And perplexing, because we had no real idea of the age of the deposits. And the associated humans—who were they? Modern or an earlier species of human, something like late *Homo erectus*?

The tools seemed too sophisticated for *Homo erectus*: surely modern humans were responsible. But then at six meters' depth, under a massive slab of fallen rock, Rokus found a small human radius (lower arm bone) that had an unusual bend to it. As our resident bone expert, he said that the radius was from an adult, but was puzzled by its size and curvature. In fact, it was so unusual that later in Bandung, where we went to get some of the bones identified, Professor Hisao Baba, an anatomist with the Natural Science Museum in Tokyo, initially thought that the radius might be from an orangutan. Baba ended up taking a cast of the radius back to Japan for more detailed analysis. Only then, with access to a large comparative collection of bones in Tokyo, did he decide that it was definitely human.

Getting a handle on the age of these deposits was now crucial. No charcoal had been recovered from below five meters' depth, so radiocarbon dating was not an option. Besides, even if we had found any charcoal, it seemed likely that the earliest deposits would exceed the capacities of that technique, which is only good for the last 40,000 years. So from Ruteng I managed to contact Bert Roberts at the University of Wollongong to ask for assistance with dating the Liang Bua deposits. Pioneering the luminescence technique in Australia, Bert and his colleagues have gathered evidence that pushes back the likely age for initial human colonization of Australia to between 60,000 and 50,000 years ago. Like a broken record, these dates had previously been stuck on 40,000 years—which it turned out was merely a laboratory artifact of the limits of radiocarbon dating.

Such long-standing debates about "when things happened" can only be resolved by accurate dating techniques. Bert and his colleagues use two main luminescence-dating techniques, which can be applied to quartz or feldspar crystals. The first technique, thermoluminescence (TL), tells you when the grains were last heated. It can be used, for instance, to date minerals in volcanic tuffs or in fired pottery. The second, known as optically stimulated luminescence (OSL), can show when mineral grains were last exposed to light. These sophisticated techniques both measure the number of trapped electrons in the crystal lattice. The electrons in the minerals are knocked free from their atoms by general background radiation and become trapped in defects in the crystal structure. The longer the exposure to background radiation, the greater the number of electrons trapped. These steadily, measurably increase with the years until the traps are suddenly emptied—by heat in the case of TL, and by light in the case of OSL—setting the clock to zero, and the trapping process begins again. To determine the age of the deposits, their background radiation is measured, then in the laboratory the number of trapped electrons is obtained by stimulating samples to emit photons, either thermally with heat or optically with light of selected wavelengths.

The luminescence techniques have many advantages over radiocarbon dating. They do not need preserved organic materials, but use grains of some of the most common minerals on earth; they can provide ages back to at least 400,000 years and in some cases more than a million years; and the principle is different from that of radiocarbon. Unlike physical dating techniques, such as potassium-argon, which date the time of mineral crystallization, OSL ages indicate the time elapsed since the mineral grains were last exposed to sunlight, and so avoids the perennial problem of dating "old" minerals that have been fortuitously buried alongside more recent fossils.

Bert and I were already trying to use OSL to obtain "burial ages" for late *Homo erectus* skulls from Sambungmacan, an alluvial terrace flanking the Solo River in Java. River terraces, and the materials they contain, are often reworked, and the Sambungmacan skulls are likely to have been deposited, eroded and redeposited a number of times before coming to rest where they were actually found in the riverbed. So their ages have always been problematic. Skulls, however, because of their bowl-like shape, capture layers of sediment. The first layer of sediment would likely be a remnant of the deposit that filled the skull soon after death, the logic being that the force required to scour out the innermost sediments would likely smash the crania. A skull, like a mini geological basin, should progressively be in-filled by younger and younger sediments. It was stratigraphy on a microscale, and we thought it was possible to determine the age of each microlayer with OSL, and then to match the age and composition of these with those of the specific terraces along the river where the *Homo erectus* remains have been found. This work is still in progress and the prospects look good.

Bert Roberts recognized the importance of the Liang Bua findings, and within a week of my call for assistance had flown from Sydney to Denpasar in Bali, then to Labuanbajo in West Flores, hired a car to Ruteng and had arrived at Liang Bua. Jack Rink, a dating specialist from McMaster University, Canada, who happened to be visiting the University of Wollongong on study leave at the time, was caught up in the excitement and came also. Jack is a specialist in a dating technique called electron spin resonance (ESR) in which age estimates are made on the basis of the number of electrons trapped in flaws in the crystals of bone or teeth, which is again proportional to background radioactivity and the length of time that the remains have been buried. This highly sophisticated technique, which has a similar age range to the luminescence techniques, measures the absorption by the electrons of microwaves in the presence of a strong magnetic field that alters the spin of the electrons. It is this alteration that is detected by the absorption of microwaves—the greater the absorption, the greater the number of electrons and the older the material. Jack had previously dated teeth from early human sites in the Middle East and

China, and had also obtained an ESR age estimate of 50,000 years for a bovid tooth from Ngandong, another alluvial terrace along the Solo River in central Java. His results were the basis of well-publicized claims by American researcher Carl Swisher that late *Homo erectus* skulls found at Ngandong were of this age or less.

Bert, itching to take sediment samples for luminescence dating, immediately grabbed a large wooden hammer and started driving plastic tubes into the walls of the excavation where we needed dates. The thwacking sound together with his various exclamations and expletives reverberated around the cave. It was the sediment right in the middle of the tube, not exposed to light, that Bert was after, but the tube sometimes hit a rock or broke and had to be extracted for another try. Given the high-tech nature of the equipment used back at the lab to measure electrons trapped in the lattice of a single crystal of a mineral, I couldn't help thinking that the technique he was using for taking samples was positively Stone Age.

While these new dating techniques have a fantastic potential to solve long-standing problems in the archaeology and paleontology of Southeast Asia, they require cross-checking, particularly when some of the techniques are still in the process of development. Accordingly, Jack Rink also took samples of the flowstones encountered during our excavation for uranium series dating—small quantities of uranium locked into the flowstone decay at a known rate to form thorium as a by-product. By accurately measuring the ratios of uranium to thorium, the age of the flowstone can be estimated very accurately. Jian-xin Zhao, a colleague at the University of Queensland, later undertook these measurements. From the original few, our network of researchers required to work on aspects of the Liang Bua excavations was growing rapidly.

As it turned out, Jack Rink's ESR results for *Stegodon* molars provided the first definite evidence for the time span in the deposits at Liang Bua. Of particular interest was an age of 74,000 years for a juvenile *Stegodon* tooth that lay just above a concentration of bone and stone artifacts. It looked like we were going to get a 100,000-year sequence from the cave with well-preserved bone from top to bottom, showing that there were major changes in the range of animals represented over this time with many surprises: *Stegodon* existed in the vicinity until at least 12,000 years ago—the youngest credible date for *Stegodon* anywhere in the world.

But most puzzling of all was the small, strangely curved hominid radius from six meters down that hinted at things to come. Our first field season at Liang Bua had got off to a promising start, but the more we discovered, the more questions arose. A different level of investigation and research was required to tackle these.

3 Planning the Project

Liang Bua is a world-class archaeological site. It has cultural deposits spanning the last 95,000 years, well-preserved bone throughout and evidence for major changes over time in human activities and animals. Our low-budget excavation in 2001 had demonstrated the site's enormous potential, but we were going to need extended periods of excavation over several years by a large labor force, and input from a range of disciplines, to unravel the cave's many stories. In turn, this was all going to require decent financial support, probably for a minimum of four years. Research might be problem driven, but is also money dependent.

Previously, Soejono and I had discussed the need in Indonesia for an archaeological project that looked at important issues right across the region and compared the cultural, faunal and environmental sequences in a number of areas—in island Southeast Asia and on the adjacent mainland. Such comparative work, including excavations at a number of deeply stratified sites, such as Liang Bua, could provide the big-picture story not seen by working in just one area. But the work had to be done systematically, had to have clearly stated aims to provide a sense of direction and cohesiveness, not the shotgun approach that historically has beleaguered Southeast Asian archaeology. Clearly stating the general aims of the project, particularly the research questions to be addressed, also meant that work could be undertaken strategically by identifying the areas, sites, disciplines, researchers and techniques most likely to provide answers.

There was no shortage of questions to tackle. The sprinkling of islands in Southeast Asia has acted like an evolutionary theater for animals—including hominids. The region also has a cultural sequence spanning a minimum of 1.5 million years; has evidence on Java for *in situ* evolution of *Homo erectus*; and has played a pivotal role in the modern human settlement of Greater Australia and Oceania.

We decided to investigate six fundamental "when, why and how" regional problems that had long remained intractable and had ramifications far beyond Indonesia.

- When did hominids first arrive?
- When and why did early hominids, such as *Homo erectus*, become extinct?
- When and how did fully modern humans first appear?
- When and why did people start cultivating plants and domesticating animals?
- When and why did technological changes, such as the introduction of ground tools, pottery and metal, occur, and what were their impacts?
- What major environmental changes occurred during the time span of hominid occupation, and what were their impacts?

Each of these questions is linked to much bigger issues—sometimes in unexpected ways. For instance, the Indonesian region has the world's warmest seawaters, which play a key role in the coupled ocean-atmosphere system and have a major influence on global climates. Our research on the local environmental context of sites, including long-term changes in vegetation, temperature and rainfall, therefore can feed into general models for climate change. In turn, such models are vital for understanding the future impacts of present-day global warming.

Underpinning many of our questions are the effects of geological and climatic processes on the evolution of Indonesia's landscape and animals. For example, the slow operation of plate tectonics and relatively rapid fluctuations in global sea level during the Quaternary (the last 2.6 million years) have been responsible for the creation and destruction of land bridge connections between Indonesian islands and the mainland. This, as you might expect, has influenced when and where animals and plants could disperse through the archipelago. Their subsequent evolution and ultimate survival or extinction in mainland or island settings have been subject to the impact of climatic changes and volcanic eruptions, with the islands of Java and Flores encapsulating the major differences in geology, biology and archaeology.

On that basis, we decided to focus on two main study areas: the Solo River Basin and adjacent limestone uplands in Java and the Soa Basin and adjacent limestone uplands in Flores. Both these areas already had proven potential for answering the questions we posed. Both have fossil sites of Lower to Late Pleistocene age, as well as limestone caves containing deeply stratified cultural deposits with well-preserved animal remains. Comparing the faunal and archaeological records obtained from very different types of sites within each area would provide better understanding of past events and processes.

On a larger scale, comparison between research results in Java and Flores would take on a much wider significance, because these islands lie on alternate sides of the Wallace Line, the primary biogeographical boundary in Southeast Asia.

The Wallace Line snakes between Bali and Lombok, then north through the Macassar Strait between Borneo and Sulawesi. At first glance, it might seem an arbitrary, inconsequential line to nowhere, but basically it marks the eastern edge of the Asian conti-

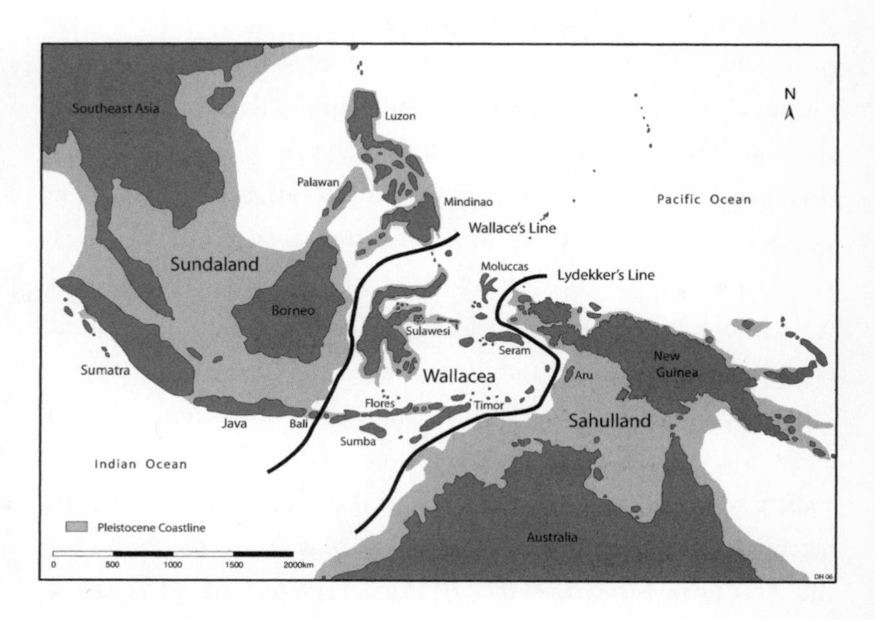

Map of Southeast Asia and the Wallace Line, which marks the boundary between continental islands to the west—such as Java—which were connected to the Asian mainland at times of low sea level, and oceanic islands to the east, which were not. Lydekker's Line marks the western edge of the Greater Australian continental shelf. (CREDIT: MIKE MORWOOD)

nental shelf, on which the seas today are so shallow that ships can anchor anywhere over the vast area between islands. At times of low sea level, this continental shelf was exposed as dry land, and islands such as Sumatra, Java, Bali and Borneo were part of the Asian mainland—and so were populated by a full range of Asian land animals. In contrast, islands farther to the east, such as Sulawesi, Lombok, Flores, Timor and the Moluccas, were separated by deep-sea barriers from both the Asian and Australian continents. Most land animals lacked the capacity to make sea crossings of such magnitude, and few managed to reach these oceanic islands.

The Wallace Line was first described by its namesake, Alfred Russel Wallace, a 19th-century English naturalist who in 1858 (with Charles Darwin) was the first to propose natural selection as the driving force for evolution. On Bali, Wallace had observed that the birds were predominantly Asian species—barbets, fruit thrushes, woodpeckers and oriental bulbuls—the same as occurred on Java immediately to the west. In contrast, on Lombok, clearly visible across a 25-kilometer-wide strait, he found that the birds were of New Guinea or Australian origin—cockatoos, honey eaters, blue-tongued lories and megapodes.

To the north of Bali, Wallace found the differences between Borneo and Sulawesi to be even more striking. On Borneo, the forests were full of different kinds of monkeys, wild cats, civets, otters, squirrels and oriental birds. On Sulawesi, to the east across the Macassar Strait, he found instead plenty of "Australian" prehensile-tailed cuscuses, honey eaters and parrots. In 1863, back in London, Wallace read a paper to the Royal Geographical Society on the geography of the Malay Archipelago—now Indonesia. He drew a red line on the map passing down the Macassar Strait and between Bali and Lombok. To the west he wrote "Indo-

Alfred Russel Wallace founded the scientific study of factors determining the distribution of plants and animals—biogeography. (PHOTO COURTESY OF HULTON ARCHIVE/STRINGER/GETTY IMAGES)

Malayan region" and to the east he wrote "Australo-Malayan region." This became known as the Wallace Line.

Wallace first identified this line on the basis of a sharp break in the proportion of Asian to Australian bird species, but more generally it marks the change from continental islands with a wide range of Asian land animals to oceanic islands with few such species. In fact, comparisons between Borneo and Sulawesi show that, while a number of Asian animals succeeded in crossing the Wallace Line, a much greater number have been stopped by it. For instance, only 13 of 59 species of shrew, 2 of 21 species of deer, 7 of 56 species of civet, and 14 of 196 species of squirrel crossed the Wallace Line. Of the primates, none of the leaf monkeys, the highly specialized proboscis monkeys, gibbons or orangutans managed the crossing—only tarsiers, macaques and humans.

Wallace realized that for new species to emerge, there had to be barriers preventing breeding between diverging populations, and he had an explanation to account for the extraordinary juxtaposition of two such radically different arrays of animals. "I believe the western part to be a separated portion of continental Asia, the eastern the fragmentary prolongation of a former Pacific continent." This sentence marks the birth of biogeography, the science of explaining where species originate and why they occur where they do. He was the first to realize that to really understand the distribution of species over the face of the earth, one had to appreciate not only the species' evolutionary history, but also the geological history of the region where they occurred.

The extent to which geology determines the dispersal, evolution and extinction of animals, including hominids, is a question that is as relevant today as it was in Wallace's time. Information on the tectonic history of Southeast Asia, including the formation, distribution, movement and character of islands, was therefore fundamental to developing our research proposal.

As imaginative as Wallace was, he would surely have been surprised to know that continents not only move up and down, but they also move laterally, rafting across the face of the globe. According to the theory of continental drift, or plate tectonics, the surface of the Earth is covered in a series of relatively thin but rigid, shell-like plates that slide over the Earth driven by gigantic convection cells of hot molten material within its interior. The often-used analogy is bits of scum riding on a slowly simmering pot of thick pea soup. An appreciation of plate tectonics is crucial because the jostling of three major plates (Pacific, Indian-Australian and Eurasian) and numerous subplates have determined the present configuration of land and sea in Southeast Asia. This intense interaction has resulted in numerous linear subduction zones—where one plate plunges beneath another—and has created volcanic arcs that formed intermittent emergent island chains. Rapid changes in topography and distribution of land and sea have provided multiple opportunities for dispersal and evolution of plants and animals.

To the north, Sulawesi and Borneo were never fully connected, with the Macassar Strait being a formidable section of the Wallace Line. Around 23 million years ago, an Australian continental fragment (eastern Sulawesi) was jammed into the subduction zone of western Sulawesi, twisting the island into its present form. Around 15 million years ago, a new subduction zone formed south of Sulawesi. This gave rise to much of eastern Indonesia, including Flores, which first emerged as a chain of volcanoes. In the southern section of the Wallace Line, Bali and Lombok may not even have existed until volcanic activity heaved them above the sea only a couple of million years ago.

Climate and associated sea level fluctuations have been another major determinant of Asia's landscape, plants and animals. Nothing could be further from the truth than to consider today's climate normal. In fact, the last time climatic conditions were similar to today's was around 125,000 years ago during the last interglacial, as the warmer periods between great expansions of glaciers are known. The present ice age is characterized by regular glacials, punctuated by interglacials (one of which we are currently in). We

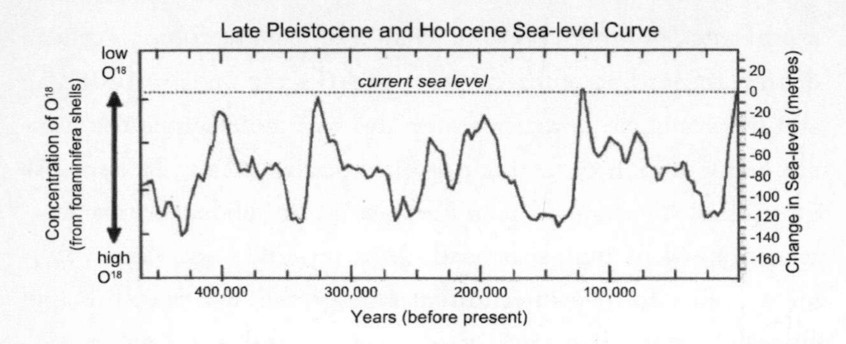

Changes in sea level over the past 800,000 years. (CREDIT: MIKE MORWOOD, AFTER IMBRIE *ET AL.*, 1984)

are, ourselves, ice age animals, having evolved during the past 2.6 million years or so, during which there have been scores of recognizable glacial–interglacial alternations. As the great continental ice sheets of the northern hemisphere and Antarctica waxed and waned, they in turn locked up and released water. Far from being in a steady state, the sea levels rose and fell, from levels similar to those of today to levels as much as 130 meters lower.

At such low sea levels, the Malay Peninsula, Borneo, Sumatra, Java and Bali were all connected by dry land, separated by a narrow channel from the composite island of Lombok and Sumbawa, which was in turn separated by a very narrow channel from the composite island made up of Komodo, Flores and Alor. To the east, New Guinea, Australia and Tasmania formed a single landmass. This vast land, "Greater Australia," was separated by as little as 70 kilometers from Timor, which was not much more than a stone's throw from Flores. At the same time, equatorial temperatures would have been cooler, and rainfall perhaps half that of today or less. Falling sea levels and a greater land area help to explain these changes, as does the differing configuration of land and water. Many currents that today enter the region would have been severed, and a climate already sucked dry by the polar ice sheets was would have been made drier.

With drier conditions between 1.8 million and 10,000 years ago, most of Southeast Asia was characterized by cool, dry, open woodland and semideciduous forest environments rather than rain forest. The fossil record of mammals on Java confirms this. As Java continued to rise, due to volcanic activity caused by the collision of the Indian-Australian and Eurasian plates, it first became an island offshore of Southeast Asia with volcanic centers surrounded by coastal swamps. Though incomplete, the land fossil record began around 2 million years ago, soon after the island was formed. The first, oldest stage, called the Satir Fauna, occurred between 2.0 and 1.5 million years ago, and included swamp-loving animals such as mastodon, a tiny hippopotamus, a giant tortoise and a deer.

Over time Java and adjacent areas of the Asian continental shelf continued to rise, and the swamps were replaced by chains of volcanoes and alluvial plains covered by open woodland and gallery forests. The scene, by a million years ago, was more in tune with present-day Africa than tropical Asia, with tigers, leopards, wild dogs, hyenas and *Homo erectus* preying on herds of *Stegodon*, hippopotamus, pig, deer, cattle and antelope. The nature and size of these animals indicate that for most of the Pleistocene, Java was a wedge of land at the edge of a continent and had a much drier climate than today's.

Of the seven different suites of mammals that can be identified in the evolutionary history of Java, only one is associated with warmer and wetter interglacial conditions. This suite, called the Punung Fauna, has recently been dated to around 125,000 years ago, and represents the most complete faunal turnover on Java in the last 1.5 million years. It only contains mammal species still living in Southeast Asia, including modern Asian elephant, Malay bear, orangutan, siamang, gibbons, monkeys, tapir, barking deer, several pigs and tiger. Gone forever, on Java at least, is *Stegodon*. The Punung Fauna also contains hominid teeth identified as modern human. Since, unlike previous stages in the faunal sequence of Java, the Punung Fauna contains no archaic animal species, it might seem likely that *Homo sapiens* also replaced *Homo erectus* in the same turnover. The problem is that the Punung Breccias predate the generally assumed point at which modern humans ventured out of Africa for the first time, around 100,000 years ago.

Throughout the Pleistocene in Southeast Asia, land animals could have walked to what are now the islands of Sumatra, Java and Borneo. Against a background of continental collision, periodic fluctuations in climate have driven dispersal and evolution. In turn, these have largely created the past and present distributions of plants and animals in Indonesia. The Wallace Line, however, remained a wide and deep barrier to the dispersal of most mammalian species of Asian origin. Even during times of low sea level, this barrier maintained its integrity. The orangutan, leopard, panther, Malay bear, Asian elephant, Malay tapir, giant pangolin, rhinoceros and banteng have never been found in Sulawesi, to the east of the Wallace Line, either living or fossil, while the islands of Nusa Tenggara, including Flores, had even fewer land mammals; only *Stegodon*, rats and hominids got to these islands on their own account.

For our planned project, the focus on major turning points in the cultural, faunal and environmental sequences meant that we would have to use analytical techniques and dating methods that would be relevant to more than one issue. For instance, luminescence methods could be used to date *Homo erectus* sites, *Homo sapiens* sites, sediments within hominid crania and early pottery. Similarly, our excavations at Liang Bua and adjacent sites in Flores could document the arrival of early hominids on the island and their extinction, the initial appearance of fully modern people, the beginnings of cultivation, the arrival of Austronesian farmers, and the first use of metal, as well as changes in the range of plants and animals over time.

For this reason, deep trench excavations in caves, such as Liang Bua (Flores) and Song Gupuh (Java), were to be "cornerstones" of the project, as were excavations at open-air fossil sites, such as those along the Solo River in Java and in the Soa Basin of Flores. The time depth represented by deep cave deposits requires use of a variety of dating techniques, while the combination of these methods would provide an opportunity to cross-check ages for consistency. Development of dating techniques would thus go hand in hand with their application.

Some of the most intractable problems of human evolution and dispersal worldwide are date debates. The dispute between proponents of the "Out of Africa," or replacement hypothesis of modern human evolution, and the advocates of the "multiregional" continuity hypothesis, is one. The former view, the dominant one among today's archaeologists, holds that anatomically modern humans first appeared around 200,000 years ago in Africa, where our species remained until 100,000 years ago, after which we swept across the world to eclipse all earlier hominid populations. This scenario, often referred to as "Out of Africa 2" because it mirrors the model for initial dispersal of hominids (i.e., Out of Africa 1), is largely based on DNA evidence from modern human populations. For instance, human DNA diversity is greatest in Africa, suggesting that populations there have had longest to accumulate differences and are therefore the oldest. Other DNA studies on isolated "relict" Negrito populations in the Andaman Islands and Thailand suggest that modern humans arrived in the region around 70,000 years ago, and that there was only a single rapid dispersal of our species into continental Southeast Asia, the Indo-Malay islands and Greater Australia, regardless of intervening sea barriers.

Available archaeological evidence supports "Out of Africa 2." At Kibish in Ethiopia, the earlier of two modern human skulls has been dated to 195,000 years ago. Diagnostic traits for modern human behavior can also be used to plot the expansion of our species. These traits include the manufacture of bone tools and stone blades, as well as use of symbols, such as rock art and personal adornments, as part of a visual communication system. They first appear sporadically at widely separated African sites from about 200,000 years ago. At Blombos Cave and Hoedjiespunt in South Africa there are bone tools and pigments associated with modern human remains from about 100,000 years ago, while at Border Cave between 80,000 and 60,000 years ago, the body of a child was covered in ochre before burial. But these sporadic occurrences do not coalesce into a full modern human behavior repertoire until later.

The oldest evidence so far for anatomically and behaviorally modern humans in Southeast Asia, from Niah Cave in Sarawak and Tabon Cave on Palawan in the Philippines, shows that we were in the region by at least 50,000 years ago. That date is supposedly supported by proxy evidence from Australia, immediately to the east, where the first human colonists were artists, made bone tools and cremated or buried their dead. The first modern humans to colonize Western Europe around 40,000 years ago had the same capabilities as the first Australians, which clearly distinguished them from the Neanderthals with whom they coexisted for at least 10,000 years before the latter became extinct.

In contrast to "Out of Africa 2" is the multiregional hypothesis, which argues that only one human species ever left Africa—*Homo erectus*—and that *Homo sapiens* arose by evolutionary changes occurring among *Homo erectus* populations throughout the Old World at the regional level; a braided stream of evolving populations, each adapting to local conditions, but still linked through consistent gene exchange. In this model the different varieties of *Homo sapiens* evolved from different regional variants of *Homo erectus* in Asia, Europe and Southeast Asia.

To tackle issues in the cultural and environmental history of Southeast Asia, our project had to be interdisciplinary in theme, incorporating not only different dating approaches, but also the study of past and present environments and animals. I therefore approached other researchers with proven track records and asked them if they wanted to join the project: Fachroel Aziz for paleontology; Soejono for his archaeological expertise; Jack Rink for ESR dating; Paul O'Sullivan for fission track dating; and Chris Turney, then at Queens University in Belfast, for radiocarbon dating. In addition, I requested funding to cover the fieldwork costs of Carol Lentfer, then at Southern Cross University in Australia, for her work on plant remains; Gert van den Bergh at the National Museum of Natural History in the Netherlands, for his analysis of excavated faunal remains; Mark Moore at the University of New England, for his study of stone artifacts; and Kira Westaway at the University of Wollongong, for her research on the environmental context of sites.

With so many people from different institutions involved, it was crucial for written agreements to be in place to spell out research and publication protocols, intellectual property rights and the procedures to cover conflict resolution should differences arise. For the archaeological aspect of the project, an Agreement for Co-operation between ARKENAS and the University of New England (UNE) was negotiated, and signed by the ARKENAS director, Dr. Haris Sukendar, and our vice chancellor, Professor Ingrid Moses. It was a general pledge of goodwill for future collaborative activities between the two institutions in a climate of open exchange and cooperation. Soejono and I were also named in the agreement as the institutional counterparts and chief investigators.

On a day-to-day level, we also negotiated specific "Implementing Arrangements" for individual projects, such as the excavations at Liang Bua and Song Gupuh. An important provision was that "specialist input from other Parties and disciplines will be on the basis of invitation after due discussion between the Chief Investigators"; that intellectual property was to be equally shared between the institutions; and that neither party could subcontract the benefit of its rights under this arrangement without the prior approval in writing of the other. For the geological aspects of the project, a similar agreement was negotiated between UNE and the Indonesian Geological Research and Development Centre (GRDC) for the geological and paleontological aspects of the project.

Bert Roberts and I spent six months writing and fine-tuning an application to the Australian Research Council (ARC) in which we were both nominated as "Chief Investigators." More than 20 researchers from a range of disciplines, as well as provision for large-scale excavations and surveys on Java and Flores, were included. This was on a different scale than that of my previous projects in Australia or in the Soa Basin of central Flores. It became clear that if the application was successful, I was to be a full-time project coordinator, not just a researcher, and would seldom get the opportunity myself to dig, survey, carry out analyses of excavated finds, or supervise the day-to-day running of associated field- and lab work. Instead, my anticipated role would be to set the general directions for the project; recruit capable and ambitious researchers with required skills; provide opportunities for experience and training of younger researchers; push things along; coordinate activities, find additional sources of funding and solve problems generally—be they political, financial, logistic or personal. Prior experience had shown that running any large project involved two very different levels of management: first, a concern with big research issues, and second, paying attention to detail. Project colleagues usually dealt with middle-range management problems.

Applications for ARC funding that year had to be received at their Canberra office by 5 p.m. on Friday, February 17, and the UNE Research Office had organized an express delivery of all applications on that day. But at 9 a.m. Bert Roberts rang to say that the Research Office at the University of Wollongong had just advised that he was ineligible to be a chief investigator on this application. He already had his allowable quota of grants, which meant our application would be rejected immediately on this technicality. Rewriting the 90-page application to exclude Bert had major implications for the costings and institutional contributions, and was going to be a lot of work. I despaired of getting a hard copy of the revised application down to Canberra from Armidale by 5 p.m. that day, but had to try. I completed the revised version just before 3 p.m., then faxed and e-mailed the various sections to Bert. He prepared 10 complete hard copies and a staff member from their research office set off on the two-hour drive from Wollongong to Canberra to deliver it to the ARC offices—after phoning ahead to make sure that someone would be there until 5 p.m. He arrived precisely at 5 p.m. and delivered the precious package of documents. Our application had been submitted on time—barely. Now we had to wait eight months to find out whether the ARC would fund the proposal.

In the meantime, from July to August 2002, we undertook a second field season at Liang Bua with excavations in a new area of the cave. This was supported by a small UNE research grant plus some top-up funding from the University of Wollongong for the participation of Bert Roberts and Jack Rink. Carol Lentfer and I also obtained a Pacific Biological Foundation grant for a project titled *Building a starch reference collection for Southeast Asia,* which enabled Carol and her Indonesian counterpart from GRDC, Netty Polhaupessy, to join the expedition to Flores. Their combined expertise would help document changes in local vegetation and people's use of plant resources over time.

Carol studies pollen, starches and phytoliths, which can survive in archaeological deposits or adhere to stone artifacts that have been used in preparation of plants. All can be diagnostic of plant species and, in the case of phytoliths (small silica structures), for different parts of a plant. With banana phytoliths, for instance, those produced by stems, leaves, and seeds of the fruit are all different. As well as microscopic plant evidence, Carol would study plant remains from sites, swamp cores and residues on stone tools. Her aim was to document changes over time in regional vegetation and the way people used plants.

Carol and Netty began by collecting plant samples for a plant starch and phytolith reference collection. For this they toured different parts of Flores, systematically collecting parts of over 300 different wild plants and crops—sometimes in the countryside along roads or the nearby Rana Mese Nature Reserve with the help of local staff from the Department of Conservation, and sometimes by purchasing produce in local town markets. After each trip, they would return with their haul to Hotel Sindha, our base in Ruteng, and sit on the veranda outside their room to patiently process piles of plant samples—each carefully bagged, documented and labeled. This painstaking work to build up the first Indonesian starch and phytolith reference collections was crucial if we were to identify microscopic plant residues. Such evidence might be microscopic, but had big implications for monitoring changes in local vegetation and human use of plants over time. The evidence would provide the basis also for understanding how, why and when farming first appeared in Southeast Asia—a fundamental question that we intended to tackle.

Today we take farming for granted, but cultivation of plants and domestication of animals was one of the greatest revolutions in human history, with worldwide repercussions for population levels and movements, present-day language distribution and socioeconomic complexity. What we call "civilization" would be impossible without farming. Plant cultivation and animal domestication, the "Neolithic revolution," began in a few nuclear centers, including the "Fertile Crescent" of the Middle East, the Valley of Mexico in Central America, New Guinea and China. The timing and sequence by which food production was then taken up in adjacent areas has all sorts of implications for later developments, as can be seen in New Guinea and Southeast Asia.

A reason why New Guinea became one of the earliest centers for agriculture in the world is because the island has an incredibly rich and varied range of plants, including the wild ancestors of some important crops, such as sugarcane, bananas, yams and taro. The oldest evidence for cultivation on the island, so far, comes from Kuk in the western highlands of Papua New Guinea, where by 9,000 years ago people were digging ditches and cultivating crops—probably taro, yam, and bananas. Carol studied phytoliths from the site and found that, unlike the familiar bananas that most of us buy from greengrocers and which only have traces of seeds in the flesh, those at Kuk had very large seeds, indicating that they were close to the wild forms. They may have been grown not for their fruit, but for their edible underground stems and trunks, and for their seeds, which could have been pierced and used in the making of necklaces.

Over time, the Kuk ditches became canal systems as food production intensified and populations increased. The onset of cultivation led to population growth and movement on New Guinea and adjacent islands in eastern Indonesia, as reflected in the presence of Papuan languages there.

The Chinese living between the Yangtze and Yellow Rivers also began experimenting with domestication early, in their case with rice and millet. It is in the nature of revolutions to spread, and the spread of Chinese agriculture was the most spectacular on Earth. It was carried to Taiwan about 7,000 years ago, whence people speaking Austronesian languages took it east to the Philippines by 5,000 years ago, and by 4,000 years ago to Borneo, Sulawesi, the Moluccas and the other islands of eastern Indonesia, including Flores.

The expansion of Austronesian speakers out of Taiwan probably began because of improvements in cultivation techniques, especially in the growing of rice and millet. Heading south, they adopted equatorial foods such as taro, breadfruit, banana, yam, sago and coconut. Charged with these, around 3,500 years ago they swept into Melanesia, along the north coast of New Guinea, and into other parts of Melanesia, as well as Micronesia and Polynesia. Their progress is marked by the initial appearance of pottery in these regions, and by distinctive items such as bark cloth beaters, spindle whorls, ground adzes and a variety of fishing equipment. And their expansion was made possible by another innovation, sailing canoes with outriggers, stable enough to sail across oceans. Ultimately, the Austronesians colonized the Pacific from mainland Southeast Asia to Easter Island, south to New Zealand, and as far west as Madagascar, in what was perhaps the greatest of global conquests; and long before the European "Age of Discovery" was even a glimmer on the horizon. Today there are about 700 Austronesian languages spread over the Asia-Pacific region.

On most islands in Southeast Asia, Austronesians appear to have swamped earlier hunter-gatherer populations—demographically, culturally and linguistically. But the contrast between what happened on Java and what happened on Flores shows that the processes of change are complex, and that insights are often gained by comparison between areas.

The beginnings of cultivation occurred relatively late on Java because, as a large, fertile continental island connected to the Asian mainland at times of low sea level, it was heavily forested and had a rich and varied range of animals, both of which favored a hunter-gatherer economy.

Evidence we later excavated from Song Gupuh, a limestone shelter in eastern Java, shows that as conditions became warmer and wetter from around 15,000 years ago, people intensified their use of the abundant animal resources and remained as hunter-gatherers until the arrival of the Austronesians only 2,600 years ago. Agricultural clearing and human population growth in Java have since resulted in the demise of hunter-gatherers, as well as the progressive loss of animal species, including elephant, rhino, tapir, bear and tiger. The island's fertility combined with irrigation agriculture now supports the highest human population density anywhere on earth, and little trace is left of the original hunter-gatherers—archaeologically, linguistically or in the features of most of the present-day occupants.

In fact, over the last 5,000 years, there appears to have been almost total linguistic replacement in island Southeast Asia. The only non-Austronesian languages to survive in the region today are the Papuan languages of Halmahera, Alor and Timor in eastern Indonesia. As in the case of New Guinea, these Papuan speakers had the economic and population capacity to largely fend off Austronesians. Indeed, it was sometimes the Austronesians who were absorbed into the Melanesian system, not the other way around. Austronesian introductions of new crops and domesticated animals, such as chickens, dogs and pigs, were simply grafted onto and helped intensify a preexisting lifestyle. The interaction between Austronesians and indigenous people on Flores over the last 4,000 years seems to have been a variation on this theme.

As a dry, oceanic island, Flores had a very limited range of land animals to support hunter-gatherers—only *Stegodon*, rats, bats and reptiles. The Liang Bua evidence shows that these animals were all-important foods for the first people using the cave, but 12,000 years ago, there was a major drop in available animals with the extinction of *Stegodon*. So from around 11,000 years ago, people began more intensive use of plants, as signaled by the appearance in the archaeological record of heavy stone implements used for pounding and grinding, starch grains on artifacts and canarium seeds. Almost certainly, these new economic activities included people bringing new types of plants to Flores, as well as cultivation. Some 8,000 years ago, people also imported the Sulawesi warty pig (*Sus celebensis*) to the island to supplement the diet.

By the time the Austronesians arrived, Papuan-speaking peoples on Flores were already long-settled farmers, with tubers, tree crops and technologies largely derived from those of nearby New Guinea, and pigs. Austronesian innovations led to the intensification of existing farming practices with the introduction of rice, millet, another type of pig (*Sus scrofa*), dog, monkey, deer, civet, porcupine and chicken. As a result, on Flores today, evidence for the original hunter-gatherers is everywhere—archaeologically, linguistically and in the features of present-day occupants. Many of the Manggarai, Ngadha and Nagakeo still resemble their Papuan ancestors rather than later Asian colonizers, and some local languages have a Papuan grammatical structure with an Austronesian vocabulary. Linguists tell me that such language mixtures, common in eastern Indonesia and western Melanesia, probably result from the fact that in areas with two distinct language groups many individuals had been bilingual.

The dating of the beginnings of cultivation and the arrival of the Austronesians with their new crops and animals are extremely important questions, and it looked like Liang Bua held some of the answers. Evidence from the site also shows innovations in local cultivation methods right up to the present day. Metal tools appeared around 2,000 years ago and would have made forest clearance easier, while the buffalo was introduced as a draft animal within the last 500 years, making it possible to prepare extensive tracts of land for wet rice cultivation. The verdant paddy fields that characterize the Wae Racang river flats in front of the cave today are, in fact, much more recent—wet rice cultivation and cash crops were only introduced to the area in the 1930s. Before that time, the Manggarai were slash-and-burn subsistence farmers who placed much greater emphasis on the growing of millet, a cereal that older people still prefer.

All this in the top 3.5 meters of the Liang Bua deposits, and we got even deeper. The previous year, six meters had seemed very deep. Now, with improvements in the construction of shoring and platforms, the same team—Soejono, Thomas Sutikna, Rokus, Wasisto, 20 Manggarai workers and I—felt comfortable excavating another sector of the site to nine meters and more. In the deeper levels, composed of layers of tuffaceous silt, we found cultural deposits that dated from about 60,000 to 12,000 years ago. They contained the remains of *Stegodon*, rat and bird, and distinctive stone artifacts including blades, awls and large flakes. Who had made them?

As the 2002 season at Liang Bua drew to a close, and already working in the red after having spent our meager budget, I began to despair that we would ever find out. Finally, in October that year, all the hard work preparing the application and the mad dash to Canberra paid off. The Australian Research Council agreed to finance our proposed project—*Astride the Wallace Line: 1.5 Million Years of Human Evolution, Dispersal, Culture and Environmental Change in Indonesia*—for the next four years. We were in business.

4 A Body of Evidence

In 2003, back in Flores with an expanded budget and field team, we could now afford to run deep excavations in two sectors of Liang Bua simultaneously. The team comprised Soejono and myself; Thomas Sutikna, Wahyu Saptomo, Jatmiko, Sri Wasisto, Rokus Awe Due and Doug Hobbs for directing aspects of the excavation; Carol Lentfer and Netty Polhaupessy for their botanical expertise; Gert van den Bergh because of his skill in identifying animal remains, especially *Stegodon*; Kerrie Grant, who was studying the excavated pottery for her Ph.D. at UNE; and Kira Westaway, who was to study the geomorphic history of the cave, take samples for luminescence dating, collect paleoenvironmental evidence, and map and date the Wae Racang River terraces.

As with all limestone caves, Liang Bua was originally formed at or below the water table, which was progressively lowered by the downcutting of the river that also formed river terraces. Kira's study and dating of deposits within the cave, including speleothems (stalagmites, stalactites and flowstones), in combination with the study and dating of river terraces outside would provide crucial evidence for the landscape and climatic history of the area.

We'd never met Kira before, but she was there waiting for us at our Hotel Sindha home base in Ruteng. Kira was young, fit, enthusiastic, feisty and inappropriately dressed in hip-hugging jeans, bare midriff and skimpy top. Manggarai Province is largely a rural, traditional, conservative Catholic area. A similar outfit next day at the site disrupted work. Whenever Kira came near, our workers stopped and gawked. Traversing the countryside surveying river terraces while dressed like that would have been risky, so I had a word with her: next day she came to the site in long military-style, baggy trousers, long-sleeved khaki-colored shirt and functional hat. Things quickly settled down and she proved a very competent and hardworking researcher. Ph.D. students can be a particularly good value on a project because they are devoted to research full-time, only have three years to complete their studies, and have yet to establish themselves. Senior colleagues with their heavy work loads, status and overcommitments can take forever.

We planned for a 12-week field season at Liang Bua and had targeted Sector I against the west wall of the cave, and Sector IV in the center, where we hoped to continue the 2001 excavation to bedrock. In his 1978 excavation of Sector I, Soejono had stopped at a sloping rock floor that he thought was bedrock, but it was actually layers of flowstone about 30 centimeters thick and later dated to between 60,000 and 50,000 years ago. To continue the excavation there, we smashed the flowstones with sledgehammers, picks and crowbars, enabling large slabs to be prised up and stacked to one side. Above the flowstones, the deposits contained the remains of present-day animals—pig, deer, dog—and modern human burials. Immediately below the flowstones were the remains of *Stegodon* and Komodo dragon with the same distinctive stone artifacts found in the deeper levels of Sector IV. The cultural deposits continued to a maximum depth of almost three meters and were underlain by more than six meters of sterile silts with striking evidence of a long sequence of cutting and infilling from water washing through the cave. Finding so much volcanic ash near the rear of a large limestone cave was unexpected. Possibly it washed in via a tunnel connecting Liang Bua with the next cave to the west, Liang Tanah, which has a funnel-like entrance into which ash could wash from surrounding areas.

About eight meters down, the sides of the Sector I excavation sprang a leak and large cracks began to appear. Even with shoring, the excavation now felt threatening and we stopped work. Soon afterward a section of the east wall weighing several tons fell—just as Kira was contemplating taking one last sediment sample from that sector. So we checked the stratigraphic drawings, pulled up the shoring timbers, and with a real sense of relief backfilled.

In Sector IV we began by removing 54 cubic meters of backfill from the previous dig, then smashing through a layer of fallen limestone slabs that formed a shelf on the west side of the square to expose *in situ* cultural deposits. This took about five days, after which more gentle archaeological excavation techniques soon yielded a human premolar tooth from 430 centimeters' depth among high concentrations of stone artifacts and *Stegodon* remains. Was the premolar going to be diagnostic? Would we now know what species of human was hunting and butchering *Stegodon*, Komodo dragon, bats and rats?

The sophistication of the stone artifacts, which included blades and awls, surely suggested modern humans were responsible, but what of the small curved radius found during the 2001 excavation? None of us on-site had sufficient anatomical expertise to pass judgment on the premolar, and it was put with other finds of special importance for later assessment.

Sector IV was a difficult excavation because of the large slabs of rockfall encountered at around six meters' depth, but eventually around nine meters down we reached the same massive silts found in Sector I, and encountered the same problems of their cracking and section collapse.

Meanwhile, outside of the cave, Carol Lentfer, wanting to look at the history of cultivation on the terraces, had begun a small excavation. She obtained permission for the work from the landholder and made good progress in her two-by-one-meter trench on the first day. However, on arriving at her excavation the following day, Carol found bamboo poles had been placed across the trench indicating that work could not proceed. Why? During the night several elders from different parts of the Manggarai region who had traditional clan connections to this block of land dreamt that the terrace had been used as a graveyard in the past and that the ancestral spirits were angry because of the disturbance.

Such problems if badly handled can easily escalate to the point where excavations at Liang Bua itself would be jeopardized—the cave also contains many burials, but these are ancient enough not to be a concern for people who now live in the area because they do not claim the cave burials as ancestors. The burial ground on the terrace was different. A solution was negotiated, and the next morning a gathering of researchers and Manggarai held a ceremony around the trench with a number of clan elders in attendance. *Adat* prayers were said and speeches of explanation and apology made. Rice was scattered on the trench and an egg wedged into the top of a bamboo stake placed at one end. A chicken had its throat slit and its blood was poured into the open trench, then its entrails were pulled out and read. We were assured that we would now have good fortune and make exciting finds in the Liang Bua excavations. These omens later proved correct. With appropriate respect and negotiation a potentially serious problem was headed off.

Inside the cave, because of the problems of collapse in Sector I, the excavation in this area was terminated early, and we turned our attention to Sector VII against the east wall of the cave. Soejono had, in the 1980s, excavated this sector, a two-by-two-meter square, to a depth of 3.5 meters, where he encountered "a layer of white beach sand," which proved to be the 12,000-year-old white tuffaceous silts. We took out his backfill quickly and excavation began in earnest. As expected, the white silts were virtually devoid of stone artifacts or bone, but underneath, we found high concentrations of *Stegodon* bone with stone artifacts, anvils and fire-cracked hearthstones. Jatmiko and Wahyu had transferred from directing work in Sector I to here, and were doing a superb job of excavating by stratigraphic layers, as well as by ten-centimeter-deep excavation units (called spits), and plotting all finds. It was evident that artifacts and animal remains consistently occurred in the higher southern end of the sector. The reason would become apparent later.

At this point, on August 7, as the work wound down, I had to return to Java to organize final payment of field allowances and other expenses associated with the excavation. Kira, Doug, Kerrie and Gert had already left, and the plan was for the work to continue for another week. Thomas Sutikna was now overall director of the excavation, while Jatmiko and Wahyu supervised the work in Sectors IV and VII, respectively. Just as I was leaving, I pointed to a concentration of *Stegodon* bones and stone artifacts just exposed in Sector VII and jokingly asked Wahyu, "When are you going to find us a premodern hominid skull to go with those?"

The trip from Flores back to Bali by the state-owned Pelni ship was a luxurious contrast to the previous ten weeks. For the princely equivalent of $80 Australian, I not only got an air-conditioned cabin for the two-day voyage, but also all meals accompanied by an orchestra. With me went the human premolar from Sector IV, wrapped in plastic and carefully packed into a film canister for protection. I had hoped that in Bandung my colleagues, Fachroel Aziz and Hisao Baba, who had looked at the strangely curved radius the year before, might be able to say whether the tooth was diagnostic and if so what species was represented. I thought it might have been from a premodern human, perhaps something like the most recent type of *Homo erectus* found on Java. Given its potential importance, the tooth would not leave my side until we knew one way or the other. Perhaps I would have a similar experience to that of Davidson Black, the discoverer of the Peking Man fossil. In 1926 he would not part with his first piece of evidence, also a tooth, of what he thought was a new ancestral human. He made a special capsule for the beautifully preserved left lower molar so he could carry it safely around his neck. Fortunately for Black, the molar was indeed diagnostic and was the basis for his claim that he had discovered a new human genus, which he called *Sinanthropus pekinensis*—although it was later renamed *Homo erectus pekinensis*.

No such luck for me. When I reached the offices of GRDC in Bandung, carefully emptied the film canister, unwrapped the precious tooth and showed it to Aziz and Baba, they were frustratingly noncommittal. Yes, the tooth was definitely human, and when compared with *Homo erectus* material held by Aziz it looked similar, but anything more they would not say. Very disappointing. Two copies of the tooth were made: one for Baba, who would take it back to Japan for further study, and one for me to take back to Australia. The original was left at ARKENAS in Jakarta.

To keep tabs on how things were going at Liang Bua, I was phoning Hotel Sindha in Ruteng every night to get a summary of progress, finds and problems. On August 10, Thomas answered the phone as if he had been sitting right on top of it. Bursting with excitement, he told me that they had just found the skeleton of a nonmodern child in Sector VII at a depth of six meters. They had found it! They had found the hominid that went with the *Stegodon* bones and artifacts. The very first year of our project was off to a flying start.

Not only had the Liang Bua excavation team found the skull of a premodern hominid, as requested, they had found a partial skeleton with some of the bones still articulated. Benyamin Tarus, one of the best Manggarai workers, had been excavating a layer of apparently sterile, sticky brown clay and, with a single scrape of his trowel, both revealed and sliced off the left brow ridge of the skull. Wahyu immediately took over and cautiously excavated some of the surrounding deposits, exposing just enough bone for Rokus to know "with 200 percent certainty" that it was a hominid skull, that on the basis of a sloping forehead, thick cranial bones and teeth with long roots it was not modern human, and that on the basis of size the remains were of a child about five years old.

Thomas provided me with daily updates as the work proceeded, describing how the lower jaw was located next to the skull, and the bones of the right leg lay next to the pelvis, with the shinbones flexed under the femur, and the kneecap still in place. By the end of the excavation, most of the leg and foot bones, fragments of the vertebral column, sacrum, scapulae, clavicle and ribs, and many finger and toe bones had been found—but no arm bones.

When first exposed, the bones had the consistency of wet blotting paper, and had to be left to dry *in situ* for two days while the team scoured chemist shops in Ruteng, buying up all available stocks of acetone nail polish remover. A weak mixture of UHU glue in acetone solvent was then applied to the bones and allowed to dry. After several applications, they were sufficiently hardened to be lifted in blocks of clay and transported back to the bone lab, Room 19 of the Hotel Sindha. For the next two weeks, Thomas and Rokus worked until late each night cleaning and further hardening the hominid remains for transport back to Jakarta. When adhering clay was removed from the skull and lower jaw, however, Rokus made a startling discovery—the teeth were worn and the third molars had erupted. This was no child, but a tiny adult; in fact, one of the smallest adult hominids ever found in the fossil record.

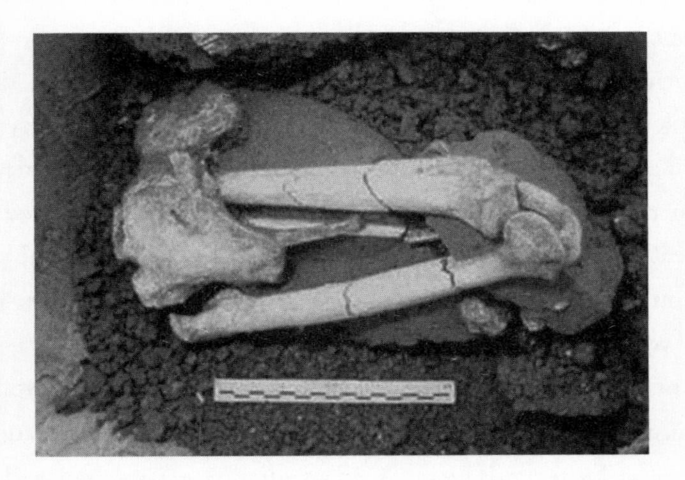

The right leg bones and pelvis of the hominid skeleton were still articulated when found, indicating that the body was partially fleshed when covered by mud at the edge of a pool against the east wall of the cave. (PHOTO: THOMAS SUTIKNA)

When it was time for the team to return to Jakarta, the bones were still very fragile. They were carefully wrapped in newspaper, packed in cardboard boxes and cradled in the laps of the team during the trip, first by ferry and then plane. It was a stunning discovery, with more than a little luck involved: even though well over 200 tons of clay, silt and rock had been excavated at Liang Bua that field season, finding a nonmodern skeleton in a two-by-two-meter square when the site floor of the cave is 2,000 square meters is still against the odds.

We do not know how the individual died. A circular lesion on the top of the skull was probably the remains of a depressed fracture from a blow—but it had healed. We knew that the corpse had not been deliberately buried; instead, examination of the stratigraphic section in Sector VII showed that it had sunk soon after death into the mud of a shallow pool of water against the east wall of the cave. This is why the skeleton was so well preserved.

I was reminded of the famous 1.6 million-year-old, almost complete *Homo ergaster* skeleton found at Lake Turkana in East Africa. Known as the Nariokotome Boy, the 11-year-old had, after dying at the edge of a lake, been quickly covered by fine muds and was thus preserved. The partial skeleton of Lucy, the diminutive australopithecine female, who lived by a lake on the edge of the lush forests in the Great Rift Valley of East Africa more than three million years ago, was preserved in very similar circumstances. Soon after death, her corpse simply sank into the soft sediments of a lake edge or stream. Heavy rains washed sand over her, and over millennia hundreds of meters of sediment built up, burying her bones deeper and deeper and turning them to stone. Later, movements of the earth's crust gradually brought her remains closer to the surface. Rains cut down the now dry earth of Ethiopia until one heavy storm washed her bones clean of sand, bringing them into the sunlight once more to be partially scattered and trampled by cattle, then found by Donald Johanson as he was walking back to his Land Rover at midday on a searing day and his glance happened to fall on a bone.

In fact, most complete skeletons of premodern hominids have been preserved only because the body of the deceased was quickly covered by muds, and so was not ripped apart, consumed, scattered, trampled or gnawed. On Flores, the bodies of the dead, if not quickly covered, would have been totally consumed by Komodo dragons. A few years ago on Komodo Island, an elderly Swiss tourist disappeared after being separated from her group. Searchers only found her camera and bloodstains.

When Thomas told me about the discovery, I was tempted to return to Flores straightaway. But my first discussions with Soejono about finding a hominid skeleton were disconcerting. He seemed determined to hand the find over unconditionally to his best friend, Professor Teuku Jacob, then aged 74, retired from the University of Gadjah Mada, but still head of the paleoanthropology lab there, and described as "the undisputed King of Palaeoanthropology" in Indonesia. Their friendship stretched back to the days of the Japanese occupation in World War II. While Soejono was risking his life to rip down a Japanese flag, Jacob was a voice on the student radio haranguing the Japanese and urging resistance.

Soejono had given all of the modern human skeletons recovered from his excavations in the upper levels of Liang Bua to Jacob—some of them more than 20 years previously. As far as I knew, none of this material had been analyzed or published. Worse still, Jacob had an international reputation for secreting away fossils. Carl Swisher, Garniss Lewis and Roger Lewin describe in their book *Java Man* the difficulty of even well-known professional archaeologists with research programs accessing Jacob's collection. They had to apply to him directly just to see the fossils, let alone touch or examine them. Only a few succeed in actually working on the fossils, and then they must wait for Jacob's final approval and organize a time with him when he alone removes the fossils from the safes where they are housed. Even Jacob's assistant, a qualified anatomist, is not permitted to study the fossils on his own. All this made me extremely nervous.

We had not put in years of planning and months of hard work and spent $100,000 on the Liang Bua excavation just to give the most important findings, resulting publications, and control of publication venues and schedules to a retired senior researcher not even connected with the project. The idea was galling. If Soejono gave Jacob the skeleton without us setting conditions for granting access, then the excavation team would be lucky to make the "Acknowledgments" of any resulting papers, let alone be credited with coauthorship. In fact, Soejono even argued that giving such credit to younger researchers was "not the Indonesian way." I replied that giving proper credit was the universal academic way, and how did he ever expect younger researchers to get decent publication track records that showed their worth?

Due to his legendary protectiveness of fossils, Jacob might not publish at all, and simply lock away one of the most fascincting archaeological finds for many years, so that no one else had access. There are precedents for this in the actions of Eugène Dubois, who in 1891 discovered "the missing link" between apes and human on the banks of the Solo River at Trinil, in East Java. Originally named *Pithecanthropus erectus*, the find was later renamed to become the type specimen for *Homo erectus*, and the most famous and controversial fossil of its time. The discipline of paleoanthropology, which Dubois pioneered, moved on, but he did not. Instead, he felt compelled to protect the unique standing of his Trinil find as the only "missing link," and refused to change his ideas in the light of further discoveries. He also had an infinite capacity for annoyance with any scientist whose viewpoint differed from his own, and a suspicious nature, verging on the paranoid. By 1900, Dubois became so rattled by criticisms of his interpretations, and attempts by other researchers to help themselves to his finds, that he placed a general embargo on them, which lasted for 40 years, until his death.

I desperately hoped that the same sort of thing would not happen today with the Liang Bua skeleton. In fact, we had gone to a lot of trouble when setting up the project to spell out the rights and obligations of participating institutions and researchers to cover conflict resolution should differences arise—hence the formal, signed Agreement for Co-operation between ARKENAS and my university, which clearly stated that the project's intellectual property, such as excavated finds, was to be equally shared between our institutions and could not be given to others without the consent of both counterparts. I was, therefore, astounded that Soejono wanted to give the hominid skeleton to Jacob, just like that and without conditions. He had other priorities and was following the rules of another game.

Anticipating "political" problems with a find of this magnitude, and needing paleoanthropological expertise, I flew from Jakarta back to Australia on August 13 to enlist the aid of my colleague Peter Brown, an experienced paleoanthropologist. Peter was a staff member at UNE, and so was covered by the intellectual property provisions of the Agreement for Co-operation. My description of the skeleton greatly interested him, but it was the copy of the hominid premolar, excavated from Sector IV and optimistically taken to Bandung for identification, that made him flush with excitement. To his experienced eye, the tooth with its multiple roots and other primitive features was hard, tangible evidence that the older levels of Liang Bua contained hominid remains that were definitely not modern human.

On the basis of this one tooth, Peter decided then and there that the Liang Bua hominid findings warranted a trip with me to Jakarta—ASAP. In preparation, I faxed Soejono a letter stating that we had already invested much time, money and effort at Liang Bua, and that we needed to be really careful that the interests of our team members and institutions were adequately safeguarded. We could discuss inviting Jacob and my colleague, Peter Brown, to provide specialist input for describing and conserving the hominid skeleton, but needed to specify an agreed time frame for such work; that the venues, timing and authorship of any associated papers should be decided by members of the project team and that the hominid remains should stay at ARKENAS.

I wanted to get back to Jakarta with Peter before Thomas Sutikna and the rest of the Liang Bua team arrived with their precious cargo. But we ran into a snag. In Jakarta the previous month, the Marriott Hotel bombing had killed 12 people and injured 150 more, so there was now an official warning from the Australian Department of Foreign Affairs about nonessential travel to Indonesia. I e-mailed Brian Stoddart, the UNE Pro–vice Chancellor for Research and International, who would need to approve the travel requisition:

> I have just returned from 3 months fieldwork in Indonesia, but circumstances have arisen which require myself and Peter Brown to return for one week. The new circumstances are—that in the last week of the excavation in Flores we discovered a complete *Homo erectus* skeleton at a depth of 5.9 metres. The skeleton is also highly unusual being an adult only one metre high with very strange teeth. It appears that hominids who first reached Flores around 840,000 years ago were then genetically isolated for several hundred thousand years and developed their own unique characteristics.
>
> This means that we have discovered a new hominid species on an Indonesian island just off the coast of Australia: a species which continued right up to the coming of the fully modern humans some 50,000 years ago. This pygmy hominid population also hunted pygmy-sized *Stegodon*.
>
> The skeleton has just been transported back to ARKENAS in Jakarta. Peter Brown and myself now plan to clean the remains, conserve them, and document them as a means of preserving Australian intellectual property rights in the find. This skeleton and our resulting publications will result in a major revision of theories concerned with hominid evolution and dispersal in this part of the world.
>
> I hope that under the circumstances UNE will agree to our brief return to Indonesia.

If this wasn't essential I didn't know what was. Brian Stoddart, no doubt with a much more pragmatic definition of *essential*, refused us permission to travel to Indonesia.

In the meantime, I had faxed Soejono another letter saying that I was returning to Jakarta to discuss publication strategies for the hominid remains recovered from Liang Bua, reminding him that charcoal samples associated with the skeleton had already been forwarded to project partner Chris Turney for state-of-the-art radiocarbon dating. Using new preparation techniques, Chris could push back the limits of this method from 40,000 to 60,000 years. I continued, saying that a quick result from this would allow us to get on with a number of scientific papers in prestigious journals such as *Nature* and *Science*. These, I emphasized, would be terrific leverage for getting more funding, and would enhance the prospects of younger professionals such as Thomas Sutikna, who needed the kudos of such coauthorship to be a serious candidate for an Australian postgraduate scholarship.

At the end of the fax, I mentioned my UNE colleague Peter Brown again, and noted that he is one of the foremost hominid specialists in East Asia, with particular expertise in the description and assessment of hominid teeth. Peter had offered to assist with describing the hominid finds; moreover, he had offered to do it immediately. This would provide the opportunity to provide specialist training to ARKENAS researchers and was very much in the spirit of our Agreement for Co-operation. I phoned Soejono the next evening to outline Peter's offer, and also emphasized that we had already invested a large amount of time, money and effort at Liang Bua, and needed to be really careful that the interests of our team members and institution were adequately safeguarded.

But it was to no avail. Soejono, acting unilaterally and against the express wishes of the younger staff who were hostile to the idea, was determined to give the Liang Bua hominid skeleton to Jacob. Fortuitously, Brian Stoddart suddenly departed on a business trip, and I was able to persuade the acting PVC, Randall

Albury, to let us go to Jakarta, by arguing that we were not going to spend long there. Randall demanded a written schedule for each day we were in Indonesia, specifying minutiae such as the length of time required to go by taxi from Jakarta airport to the central railway station and the distance from the offices of ARKENAS to the Jakarta central business district. It was one of the most creative bits of schedule writing I've ever done.

On Monday, September 21, 2003, Peter and I flew to Jakarta. Our arrival coincided with the delayed return of Thomas, Jatmiko, Wahyu and Wasisto, who with the Liang Bua remains had stopped over in Bali for a well-earned break. The next day we met Soejono and the just-returned excavation team at the ARKENAS offices. The remains of the skeleton were taken into the conference room, spread out on tables there and carefully unwrapped.

I observed Peter's reaction. He paled and went quiet when he saw the skull, lower jaw and femurs. It was a dead giveaway. In fact, no one spoke much, but there was a real sense of suppressed excitement and even disbelief. Peter refused to discuss the material at all until decisions had been made about how the skeleton was to be analyzed and published—and by whom. It was clear that Soejono had already made commitments, and he insisted that the remains be given to Jacob right away. While polite to Peter, he was actually seething underneath at another paleoanthropologist intruding on the scene. Wahyu had never seen him so angry. It turned out, however, that Peter's presence was our lifeline. Without him standing there awkwardly, but prepared to clean, conserve and analyze the remains right away, we would have had no bargaining position at all. Apparently, Peter did not enjoy his first Jakarta experience.

With Peter and the younger Indonesian researchers looking on and saying very little, Soejono and I had a long, acrimonious discussion. He was certainly not used to being challenged in this way, but given what was at stake, I felt there was no choice, and was finally able to use the provisions of the UNE/ARKENAS agreement to negotiate a proposal acceptable to all. We would submit two papers to the most prestigious scientific journal in the world, *Nature*, where our findings and claims would be subject to a stringent review process by experts in the field. For the first paper, on the hominid skeleton and its taxonomy, Peter Brown and Thomas Sutikna would be lead authors, while other members of the excavation team—Soejono, Wahyu, Jatmiko Rokus Awe Due and I—would be coauthors. For the second paper, on the age and chronological context of the skeleton, Soejono and I would be lead authors, while all other researchers who had contributed—an eventual total of 17—would be coauthors. We also agreed to place an embargo on all information concerning the find, because premature release or publication of photographs in any media would jeopardize chances of getting published in *Nature*, which would be ruthless enough to reject a find even of this magnitude if it had been leaked to anyone else beforehand.

It could easily have gone the other way or bogged down in stalemate. If the skeleton had disappeared into the bowels of the paleoanthropology lab at the University of Gadjah Mada, like so many previous finds from Liang Bua, who would now know of its existence or real significance?

The next day, Peter began work conserving and studying the Liang Bua hominid remains, assisted by Thomas, Wahyu, Rokus, Jatmiko and me. The skull had to be cleaned of adhering clay both inside and out, and then further consolidated to allow measurements to be taken. This was a painstaking business involving use of dental probes, picks and paintbrushes. It took three days to remove the sediment filling the skull before we could measure its cranial capacity, an approximation of the brain size. For this, Peter had brought a decanting flask and a parcel of mustard seeds from Australia. He carefully blocked all the holes in the skull, bar one, and then poured in the mustard seeds. As primitive as it sounds, it is an effective technique.

Three times, Peter, assisted by Jatmiko, filled the cranium, then emptied the seeds out and measured their volume, while the rest of us watched with bated breath. On the basis of the outside skull dimension, Peter had been anticipating a cranial capacity of around 600 cubic centimeters (cc), a line-in-the-sand measurement at which the find could be regarded as "human." Instead, because the bones of the skull vault were much thicker than expected, the measurement was an astonishingly small 380 cc. They measured and

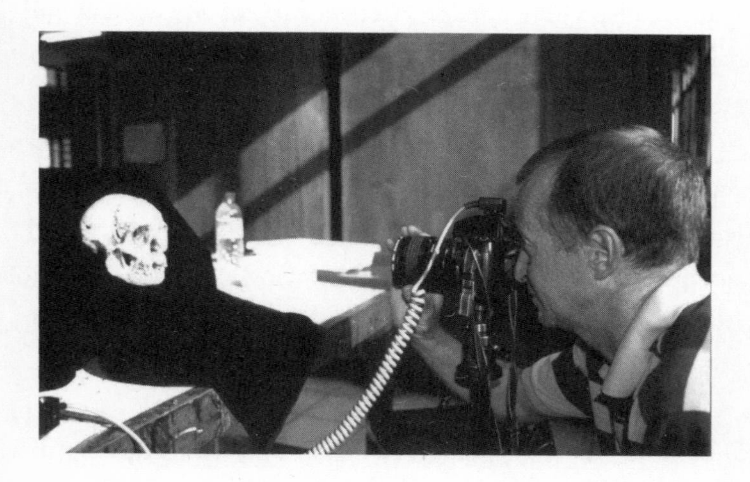

Peter Brown photographing the LB1 skull in the collection room at ARKENAS, September 25, 2003. This work only began after a sometimes-acrimonious discussion between Soejono and me, while the rest of the excavation team watched anxiously. We finally decided to try for two papers in Nature *that later appeared in October 2004.* (PHOTO: MIKE MORWOOD)

remeasured the cranial capacity. There was no mistake. It was consistently, stubbornly below the defining capacity of the genus *Homo*. Peter mumbled that this hominid would be flat out chewing nuts and leaves. How could it have made stone artifacts or hunted *Stegodon*?

The definition of genus *Homo* has always been difficult because it is closely tied to the concept of "being human." In fact, the genus was first designated by Carl Linnaeus, the Father of Taxonomy, in 1758, when scientifically naming our species, *Homo sapiens*—meaning "wise man." Big brains defined being human. It was such an obvious fact that no one really questioned it. After all, we were part of God's work. That was until Darwin unleashed evolution on the world with *The Origin of Species* in 1859. One sentence tentatively mentioned, almost whispered, right at the end of that treatise—"Light will be thrown on the origin of man and his history"—suggesting that humans too evolved, caused a storm. He may as well have shouted it in God's face. It took decades for the fury to die down and for it to be safe enough to ask the question: How small a brain can you have and still be human? Even today, the answer to that question is that we really don't know.

Modern humans have an average brain size of around 1,200 cc, with normal variation between 1,000 and 2,000 cc. Some present-day individuals can be afflicted with a condition known as microcephalia, in which the size of the head and brain are well below normal, as a result of genetic abnormalities, disease, or exposure to toxic substances at an early stage of development. However, cases of extreme microcephalia, in which the brain is below 600 cc, are extremely rare, and result in mental retardation and other problems. The affected individuals usually cannot speak, plan, or perform complex actions, and generally do not survive to adulthood.

In the first half of the 20th century, Sir Arthur Keith, then the doyen of British anatomy, thought that the answer to the question How small a brain can you have and still be human? was "The smallest brain that we have." His figure was 750 cc, and it became widely accepted. In the 1960s and 1970s, the famous Leakey family decreased the acceptable brain size for genus *Homo* first down to 600 cc and then further to 500 cc when describing *Homo habilis*, the earliest known species of human. Could we take the smallest acceptable human brain size down to 380 cc? This would be the smallest known brain capacity for any hominid species—even australopithecines like Lucy—and overlapped with that of chimpanzees, which have brain sizes ranging from 300 to 500 cc.

After a week of working from early in the morning until late at night, the first pass of the Liang Bua hominid analysis had been completed, and we left Jakarta by train for Bandung, as we had promised Randall Albury. Peter, who had formally designated the skeleton as LB1 (Liang Bua 1), now wanted to compare some of its traits, especially the structure of the inner ear bones, with those of the reference collection of chimpanzee, gorilla and orangutan remains held at GRDC. Fachroel Aziz and his Dutch colleagues, John de Vos and Paul Storm, were there, and were keen to know what had turned up on Flores that was obviously exciting us so much. The primate remains were assembled for scrutiny, and Aziz also took out some of his precious *Homo erectus* skulls from his wall safe—GRDC is the legal repository of fossil hominid finds in Indonesia, and Aziz is curator for their remarkable collection. Careful comparisons established that the Liang Bua hominid, now identified as an adult female, did not have an ear bone structure like that of modern humans, Indonesian *Homo erectus* or orangutan. Instead, it most resembled those of African great apes!

Despite the difficulties, things had gone well. It was unanimously agreed that we try to get two papers on LB1 published in *Nature*. Peter had been allowed to undertake analysis of the major parts of the skeleton, and we had negotiated the authorship conditions for his input, which ensured that our institutions, project and team members would be given proper credit. Finally, everyone concerned knew that information on the find was to be embargoed. Premature release of information or publication of photographs in any media would jeopardize our chances of getting properly published in a peer-reviewed journal. This was essential for establishing the credibility of the find and its significance. Meanwhile, Thomas Sutikna was to continue cleaning and conserving LB1, which had been stored in a dilapidated, locked filing cabinet in Soejono's office. Thomas also acted as gate-keeper: he had the only key to the cabinet, restricted access to the remains, and let no one else take photographs or measurements. That is why the embargo held for over a year, but it was a constant worry for the whole time. We felt the sooner we got the papers written, submitted and published, the better.

We returned to Jakarta the next day, September 30, and that evening flew back to Australia. On the flight back, Peter's checked-in luggage with all his photographic equipment, rolls of film and notebooks containing the LB1 data went astray.

"Bugger" was all he said.

5

The Devil of Dogma

After a worrying few days, Peter's bags turned up. And, after he had further examined his notes and photographs, he made up his mind that on the basis of the cranial capacity and other primitive features, LB1 could not be included in the genus *Homo*—so she was not human, but more like an australopithecine. He suggested an entirely new genus and species name, *Sundanthropus tegakensis* (erect ape-man from the Sunda area). Surprised by this choice of name, I commented that only Indonesian- or Malay-speaking peoples would know or remember the term *tegakensis*, so we decided to substitute the name *floresianus*.

In addition, there were good reasons for not referring the species to a new genus. Far better to change what it means to be accepted as *Homo* than to invent a new genus name. This is precisely what Louis Leakey and two colleagues had done in 1964 when they announced their discovery of *Homo habilis* and in doing so dropped the accepted cranial capacity for genus *Homo* to 600 cc. For this they had copped a lot of strident criticism, but they persevered and eventually won out. Selecting the right name for the species was important scientifically and politically, to ensure that

LB1 was not regarded as just some Southeast Asian oddity of little relevance to the understanding of hominid evolution and dispersal generally. But Peter was adamant in his decision to allocate the specimen to a new genus. He said that Bernard Wood, a leader in the field of paleoanthropology, had written definitively on the traits that define us as human, and had set a minimum 500 cc cranial capacity for *Homo*. So LB1 could not be *Homo*.

Actually, LB1 has characteristics of both *Homo* and *Australopithecus*. Her skull, while tiny and distinctive, is similar to that of *Homo erectus*, long, low and relatively broad with the maximum width being low down. The chief difference is its size—it is much smaller, and the brow ridges, while pronounced, arch over the eye sockets and do not form a straight bar projecting out to the sides, as in Indonesian *Homo erectus*. Inside the skull, the brain of LB1 was tiny at 380 cc, but as an endocast later showed, apart from its size, it was also similar in shape to that of *Homo erectus*—except that it had enlarged frontal and temporal lobes—precisely those areas concerned with cognition and planning. This is not what we expected.

Like the australopithecines, LB1 has canine teeth with pronounced roots that form prominent thickening of the facial bone right up to the nose. However, she does not have the projecting muzzlelike prognathism, or relatively long face, of australopithecines, and the overall shape of the face is similar to that of other members of the genus *Homo,* as are the parabola-shaped dental arches, which are a feature used to identify the earliest evidence for genus *Homo* in the African fossil record.

The lower jaw lacks a chin. Instead, there are bony supporting shelves at the midpoint of the inner curve of the lower jaw (the symphysis), a trait found in australopithecines, and the bone is thickened at that point. One of the hominid lower jaws excavated at the 1.8 million-year-old Dmanisi site in the Republic of Georgia

DOUGLAS COLLEGE LIBRARY

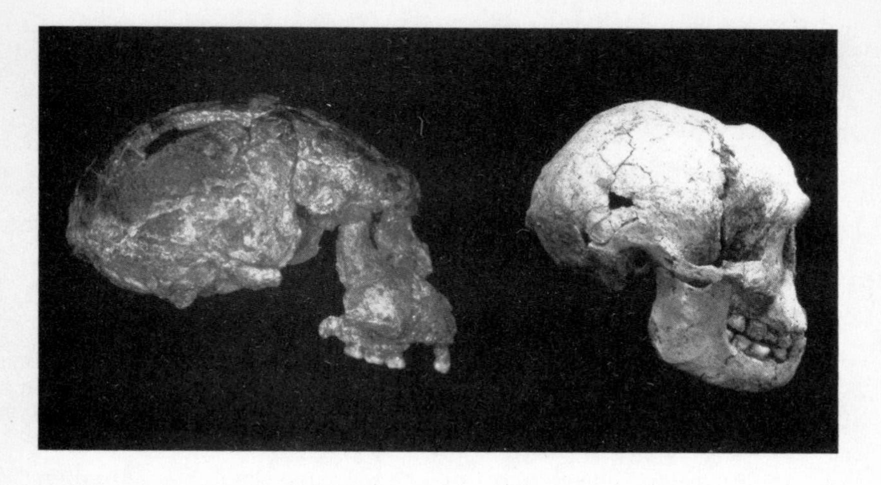

Comparative side views of the skulls of LB1 Homo floresiensis *and Sangiran 17, the most complete* Homo erectus *specimen from Sangiran in Central Java. Despite differences in size, there are a number of specific similarities. Both are long skulls and low.* (PHOTO: PETER BROWN)

has similar features. The relative sizes of teeth in the LB1 lower jaw are typical of genus *Homo*: the third molar is smaller than the first and second molars, which are approximately equal in size. In contrast, the third molar of *Australopithecus* is larger. However, the crowns and twin roots of LB1's premolars are similar to those of australopithecines and early *Homos* including some of the Dmanisi specimens.

The most immediately striking feature of LB1 is her extremely small size. For instance, the femurs are a mere 28 centimeters in length—shorter than those of Lucy, and equal to the smallest femur estimate for *Homo habilis*. Applying formulae derived from modern human pygmies to estimate stature, we find the femurs of LB1 suggest that she was about 106 centimeters in height. But since she lacks a forehead, this calculated height is likely to be an overestimate.

The arm bones of LB1, which we excavated later in 2004, pro-

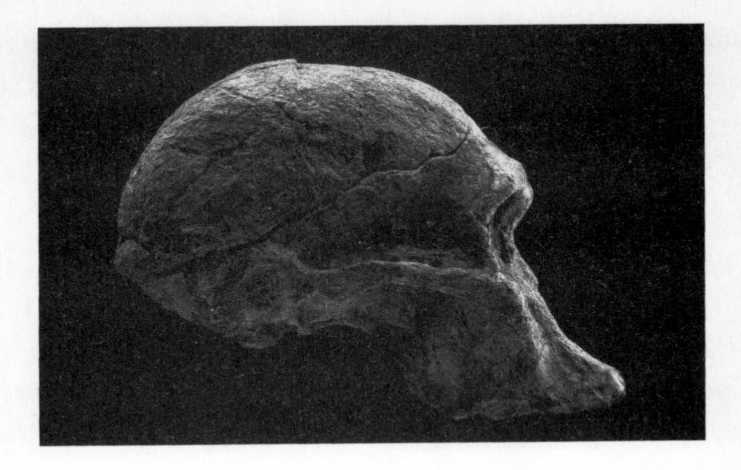

Side view of the skull of Australopithecus africanus*—the man-ape that lived in South Africa around 2.5 million years ago. This skull shows the projecting, muzzle-shaped, lower face typical of* australopithecines, *which results in a more V-shaped dental arch with straighter tooth rows. In contrast, LB1 has the flatter face and parabolic dental arch typical of genus* Homo. (PHOTO: DAVID BRILL, COURTESY OF TRANSVAAL MUSEUM)

vided further useful information. Although short, LB1's limb bones are all much thicker for their length than modern humans; in fact, almost identical to those of chimpanzees (and Lucy). Despite her midget size, she was muscularly powerful and her body weight is probably best estimated by comparison with chimpanzees rather than modern humans—meaning that she probably weighed about 30 kilograms. The evident strength of LB1's forearms indicates that she may have been very adept at tree climbing, perhaps to collect food, find a place to sleep or escape predators. The bones of the hands and feet are also very curved, again probably indicative of tree climbing.

The recovery of LB1's collarbones and right upper arm (humerus), as well as the shoulder blade of another tiny individual, was a real bonus. Shoulder joints contain a lot of information about locomotion and dexterity, but are not common in the fossil record because these bones are relatively fragile and are surrounded by meaty muscles that are attractive to carnivores. They are usually only preserved in relatively complete skeletons, such as Lucy, the Nariokotome Boy and LB1.

It was soon apparent that her shoulder anatomy differed from ours in a number of ways. For instance, in modern humans the top (or head) of the humerus is angled between 145 and 165 degrees to the plane of the elbow joint. As a result, the inside of your elbows face slightly forward when you stand straight with your hands by your sides, and the elbows can be bent to enable you to handle things in front of your body. But in LB1 the head of the humerus was only twisted 110 degrees, which is the norm for gibbons and macaque monkeys. This would seemingly have restricted her ability to work with both hands together in front.

More detailed study by Susan Larson, a researcher from New York's Stony Brook University specializing in primate shoulder anatomy, however, found that the collarbones of LB1 were relatively short. As a result, the shoulder blades would have been more shrugged forward, so that she could still bend her elbows and use her hands as we do, but with limited capacity for throwing. Nor would she have been capable of endurance running, which requires backward and forward movement of arms along with rotation of the chest to counterbalance the movement of the legs. When Susan undertook comparative analysis of an early hominid, the Nariokotome *Homo ergaster* skeleton, she was surprised to find that he has exactly the same humerus anatomy as LB1. Both specimens provide evidence for an earlier, previously unsuspected, hominid shoulder configuration: one that seems transitional between those of apes and modern humans.

Her skeleton contains more clues for how LB1 would have looked and moved. While her pelvis showed that she was definitely a bipedal walker—a defining trait for hominids—it was relatively broad and more flared than ours. This would have given her a potbellied figure, again more characteristic of australopithecines than *Homo.* The relative length of her arms and legs is also very telling when compared with that of apes and other hominids. For instance, in chimpanzees the humerus is about the same length as the femur, yielding a Humerofemoral Index of 100 percent, which compares with an apelike 85 percent for *Australopithecus afarensis* (Lucy), 85 percent for LB1 and 70 percent for modern humans. Although LB1 had almost exactly the same body proportions and height as *Australopithecus afarensis,* Bill Jungers, a researcher from Stony Brook University specializing in hominid postcranial bones, has observed that LB1 has short legs relative to her height, whereas Lucy has particularly long arms relative to hers. In this case, similarities in body limb proportions between two tiny hominids such as Lucy and LB1 may reflect evolutionary convergence rather than close family connections—another reminder that hominid morphology seems to have been much more changeable than generally believed and that simple quantitative comparisons between species as a basis for working out family relationships may be misleading.

The feet of LB1 are of particular interest. They are extremely long compared with the leg bones—over 60 percent of femur length. Bill Jungers concludes that long feet may have been means for increasing the stride length to compensate for short legs—an adaptation found in no other hominids, and one in which speed may have been sacrificed for energy conservation and stability. In addition, some of the bones along the outside of the foot allow greater flexibility between joints than is the case with modern

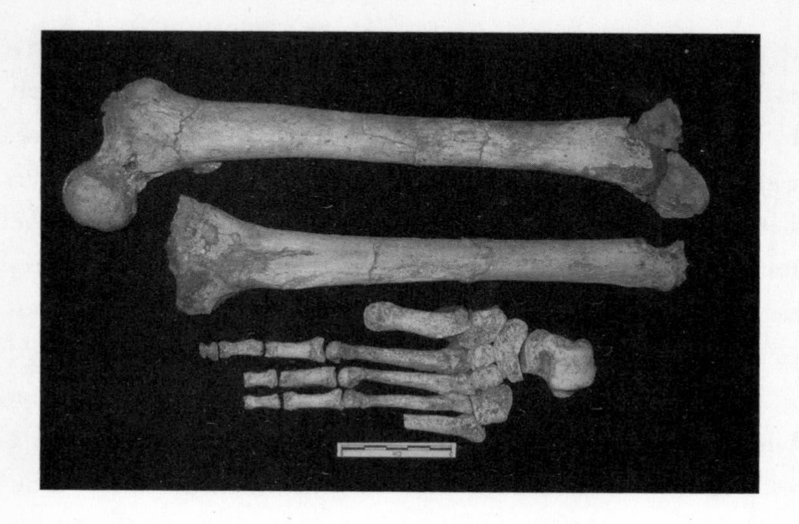

LB1 lower limb bones. These are extremely short in comparison to the body and arms, but also thick and robust. The foot is also very long, which would have increased the length of the stride. The curved phalanges (toes) may indicate a more arboreal lifestyle. (PHOTO: BILL JUNGERS)

humans. This is a more primitive condition that has previously only been documented in *Homo habilis*!

At Liang Bua we eventually found remains from at least 12 other individuals similar to LB1, but in all other cases they were found to be shorter. For instance, a tibia (lower leg bone) from another adult (LB8) indicated a height of only 101 centimeters on the basis of African pygmy data—again, this is probably an overestimate. Even though these individuals represented a time span of more than 80,000 years (i.e., 95,000 to 12,000 years ago), all had the same strange traits, including very thick limb bones, and relatively long arms and short legs.

A second lower jaw excavated at Liang Bua is from another smaller adult (LB6), and is very similar to that of LB1 in size, shape and anatomical details, despite the fact that it came from occupation levels about 3,000 years younger. It provides another line of evidence that LB1 was not a pathologically deformed individual, but is representative of a long-term hominid population with a unique combination of traits. The second jaw also showed that there was individual variation within the population. It is more robust and has a more V-shaped dental arch with straighter tooth rows than that of LB1. In fact, it is eerily reminiscent of australopithecine lower jaws.

I argued with Peter that the amalgam of traits we found on Liang Bua hominids most likely meant that the original population in Asia was very early in the line of *Homo*, possibly a transitional *Australopithecus/Homo* population, such as *Homo habilis* or *Homo rudolfensis*—which both appeared in East Africa around two million years ago. These constitute the earliest known species in genus *Homo*. Some researchers argue that because of their short stature, small brains and apelike body proportions, both these species should be referred to the earlier australopithecine genus, namely as *Australopithecus habilis* and *Australopithecus rudolfensis*, but you might expect such taxonomic blurring and uncertainty at times of transition. Was LB1 another example of taxonomic blurring?

LB1's brain size was well below the previously accepted range for inclusion in the genus *Homo*, but such definitions are not fixed absolutes, and she was definitely no chimpanzee. Did we want to deny the humanity of the little people at Liang Bua who had made stone tools, used fire and hunted mini-*Stegodon* communally? I was treading on dangerous ground here because the Liang Bua evidence challenged a fundamental belief—in this case, what a human could be. That's never an easy thing to do even if all the evidence is in your favor. There are many precedents for reactionary conservatism in the history of science.

In the early 20th century, for instance, most researchers still expected that species ancestral to humans would have had big brains like ours. But for anyone who was looking, the several Neanderthal fossils already known, together with Dubois' *Pithecanthropus* fossils from Java, which were clearly more primitive than those of Neanderthals, pointed to an uncomfortable conclusion: ancestral man had a small brain, and it got smaller and more apelike in structure the farther back you went. In contrast, early human teeth and jaws were not like those of apes, and resembled those of modern people. It was the reverse of what had been generally anticipated: that an increase in brain size occurred early in our evolutionary history to distinguish us from other animals. Increasing numbers of human fossil finds required some contortions of logic to explain away the evidence, but the standard ploy was to interpret differences as being due to pathology or a disease such as rickets rather than being indicative of a different species.

Then, in 1912, something that fitted right in with the dominant view finally did turn up—and it proved 40 years later, after much fanfare, to be a fake! In the nondescript town of Piltdown, East Sussex, England, Charles Dawson, a solicitor and amateur archaeologist, found a collection of fossils—pieces of human skull, monkeylike jaws and a few teeth—in some excavation works. He took the fossils to one of the foremost men in British science, Sir Arthur Smith Woodward, the keeper of the Department of Geology in the South Kensington Museum. He immediately declared that they belonged to an early human species. "Piltdown Man" had the "expected" mix of features—a blending of human and ape with the noble brow of *Homo sapiens*, and an ape lower jaw—which lent it credibility against the other lowbrow fossils.

Piltdown Man's lofty dome satisfied the prejudice that men are glorious, unique and intelligent. The excitement caused by these finds was great, and the English newspapers took up the cry that "Dawn Man" had been found in the south of England—even though it turned the evidence of all other human fossils upside down. This fact was not lost on Ales Hrdlicka from the Smithsonian Institute, who correctly claimed that the lower jaw came from an ape, and the cranium from a modern human. He said that the bones from Piltdown could not be right if the bones of *Pithecanthropus erectus* and Neanderthals were also right.

Over time and against new discoveries, Piltdown Man became marginalized. Then, in 1953, at an international conference of paleontologists, the world cache of fossil humans was examined. Against such a line-up, Piltdown Man simply did not fit. Once the possibility had been allowed for, it became ridiculously easy to see that the finds were fake. For a start, the teeth had been filed to fit a misfitting lower jaw—but the wear was the wrong way around. There were never any fossils at the Piltdown quarry. Instead, the quarry had been salted from time to time with bones from a variety of sources and treated to make them appear to be genuinely ancient: a medieval human skull, an orangutan lower jaw from Sarawak, a Pleistocene chimpanzee fossil canine, a fossil elephant molar from Tunisia, and a fossil hippopotamus tooth from Malta or Sicily.

It was a brilliant fraud, which had escaped notice simply because it satisfied the dominant—but flawed—paradigm. It suited British vanity and it was testament to the political naïveté of science. To this day, no one knows who the perpetrator was, although Dawson is the chief suspect.

LB1 definitely is not derived from a *Homo erectus*, as found in Java or China. She had *Australopithecus* stature, brain size, and body proportions; *Homo* dentition and face structure; and some totally unique traits. We did not know where she was going to fit in the general scheme of hominid evolution and progress, but I believed that referring her to a new genus would bury the connection with early *Homo* and its australopithecine ancestor.

The first major discovery by Louis Leakey, *Zinjanthropus boisei,* was nearly consigned to marginality when allocated to a new genus, which obscured its relationship with other hominids. The species was later rechristened *Australopithecus boisei*, then reallocated to the robust australopithecine genus as *Paranthropus boisei*. Even if we didn't know where exactly she fitted, I didn't want this to happen to LB1, but it seemed like it might. Peter's main concern was that her brain size was simply too small to be considered *Homo*. But should size matter? I thought. Surely, it is more a question of cognitive capabilities—and there was abundant evidence that the Liang Bua hominids were smart. While they did not make adornments, paint, or bury their dead, they made use of fire, and were handy with scrapers, anvils, points and assorted stone implements.

I let my mind drift into the past to try to capture the emotions and feelings of these tiny humans who had once been alive, sheltering in Liang Bua, bringing in hunted game and vegetables, or bundles of firewood to be carefully used for cooking, warmth and light. I could see them in my mind's eye carrying in river cobbles for the hearths, which emerged intact from the sediment of the cave; selecting stones for anvils; squatting to make tools for the butchering of Komodo dragon, *Stegodon* and giant rat. I imagined them discarding the smashed remains of the skulls and the charred long bones, leaving the tools smeared with the fat, blood and hair that we would later find and identify; imagined seeing them sitting quietly while concentrating on some woodwork, or communicating while repairing or hafting implements, or preparing food; at least 4,500 generations of people sleeping, socializing, playing, fighting, making love, giving birth, rearing children and dying.

What had Liang Bua and its environs looked like when LB1 was alive? The range of animals represented in the Liang Bua deposits suggests that during the Pleistocene, when conditions were colder and drier, there were areas of grassland and open forest in the vicinity, while rain forest probably fringed the river itself. For instance, we recovered the remains of a bearded vulture, never previously reported from Southeast Asia—its present distribution includes the Himalayas and other high mountain ranges of Central Asia, and for good reason: this scavenging bird locates carcasses by sight, and needs areas of open country to survive. In contrast, young Komodo dragons need trees, where they spend a lot of time avoiding being eaten by their cannibalistic elders. The Liang Bua hominids would have walked across open country between areas of forest, but probably had to take to the trees on occasion to avoid being eaten by Komodos, which often sneak up on sleeping animals.

At the time, the interior of Liang Bua was also very different from today: water entered the cave from the rear and from a passage on the west side connected to another cave from which a stream issued to intermittently cut, infill and recut channels across the cave floor. Against the east wall, where we found LB1, there was a standing pool of water. Perhaps she lay down to die at the water's edge, sinking in the shallows while still partially fleshed, to be rapidly covered by fine silts. Adjacent to the pool, toward the back of the main chamber, there was higher ground, dry and level. Here we found a series of living floors, superimposed one on the other over thousands of years, and densely littered with stone cores, flaking debris, retouched tools, flat pebble anvils, fire-reddened hearth stones and animal remains. The artifacts were left exactly as last placed. You could actually crouch at the exact spot where someone knapped a split cobble, leaving a halo of simple flakes.

Our excavations found the remains of *Stegodon*, Komodo dragon, rat, bat and bird, with clear evidence that some animals were butchered on-site. The skulls of cranelike birds, Komodo dragons and *Stegodon* had been smashed; some bones were charred or had cut marks; different sorts of hair were also found in the excavated deposits—some thick and black, others thin and pale. Occasionally, hominid remains were found on the same living floors—for instance, a radius, scapula, ulna and jaw found together about five meters down and dated to around 15,000 years ago—but in no case was there any evidence that hominids had been items of diet. No hominid bones had evidence of cut marks or charring.

The stone artifact technology of the little hominids was distinct from that associated with modern humans in higher levels of the deposits. Cobbles of volcanic rock, or chert, were obtained from the nearby riverbed, and split on flat-topped anvil stones. Flakes were then struck off from the margins of the split cobbles with small hammerstones. We found larger flakes that were probably made outside the cave, at the actual source of stone, and then carried in to be used or further worked. Some flakes were modified by retouch, but most were not. Looking closely at some flakes, I could see heavy edge damage, which indicates that they were used for the working of wood. The concave working edges on many stone tools also indicate that wooden shafts were being worked—maybe as digging sticks, spears and clubs.

As well as simple flakes, there are more formal stone artifact types in the Pleistocene levels of Liang Bua. These include points and blades, some of which were probably used as spearheads and knives for hunting and butchery, small blades that may have been hafted as spear barbs, and awls used for making holes in wood or hides. These tools are found only with evidence of *Stegodon*. The same range of tools and methods for flaking stone are found at Mata Menge and other stratified sites in the Soa Basin, the oldest of which we had dated to about 840,000 years ago. These sites also contain the fossilized remains of the large-bodied *Stegodon florensis*, which, according to Gert van den Bergh, was directly ancestral to the dwarfed *Stegodon florensis insularis* found at Liang Bua—with the latter having shrunk 30 percent in body size over the intervening time.

The exact match between stone artifacts from Mata Menge and the Pleistocene levels of Liang Bua appears to be a remarkable example of hominid technological and genetic continuity spanning some 830,000 years. Yet some still argue that the stone artifacts at Liang Bua are too sophisticated to have been made by the tiny, small-brained hominids, whose remains occur in the same occupation levels. Instead, they say that modern humans lurking in the archaeological shadows must have been responsible. Maybe, but the Mata Menge evidence says otherwise, while the distinctive stone artifact technology associated with LB1 and her kin continues from the oldest cultural levels at the site, until the disappearance of *Stegodon* and *Homo floresiensis*, immediately below the "white" tuffaceous silts blasted from a volcanic eruption 12,000 years ago, which threw a death shroud across the local landscape.

Even more convincing is the fact that there is no skeletal evidence for modern humans at Liang Bua below those white tuffaceous silts; nor the least hint of symbolic behavior, such as pigments, art, adornments or formal disposal of the dead, which are core characteristics of all modern human cultures. In contrast, there are lots of modern human skeletal remains, burials, pigments and ornaments in deposits above the white silts, where evidence for tiny small-brained hominids is conspicuously lacking.

Did modern humans and *Homo floresiensis* ever meet at Liang Bua? Not on the basis of current evidence, but we've hardly scratched the surface in our archaeological investigations. We do know that for some 80,000 years, Liang Bua served intermittently as a home base—a place to which groups of tiny humans consistently returned, bearing foods and materials from the surrounding countryside. Even very light use of the cave by a group of, say, 10 people for 15 days a year means that it may have been occupied for over 12 million person/days during the time that *Homo floresiensis* is known to have been there. Only a very small percentage of this use would have left material traces, and much less would have survived, but the site is still very rich in evidence. So far only about 1 percent of its deposits have been excavated, and our knowledge about the archaeology of Liang Bua, and more generally, Flores, is very preliminary.

Wider afield, there is no equivalent information at all concerning early hominid occupation of adjacent islands, such as Sulawesi, Sumbawa, Timor and Sumba. All are exciting prospects for future research. If early hominid populations reached and survived long-term on these islands, they would have been subject to some of the same evolutionary pressures evident in *Homo floresiensis*. Future discovery of other hominid species in the region is a near certainty—and each will have its own story to tell.

Jack Karadada, a Wunambal Aboriginal elder and traditional owner of part of the Kimberley coastline in the northwest corner of Australia. Jack's family have been in Australia for some 2,500 generations, but their ancestors were probably Indonesians. *Photo: Mike Morwood.*

Mounted Ngadha hunters in the Soa Basin, armed with parangs, spears and harpoons, in search of pig and deer. *Photo: Mike Morwood.*

A painting at the Ledalero Catholic Seminary Museum depicting the Soa Basin as Father Theodor Verhoeven envisaged it 750,000 years ago. His excavations at Mata Menge and Boa Lesa in 1963 demonstrated that early humans had reached the island by that time, and they had shared the island with large Stegodon. *Courtesy: Theodor Verhoeven.*

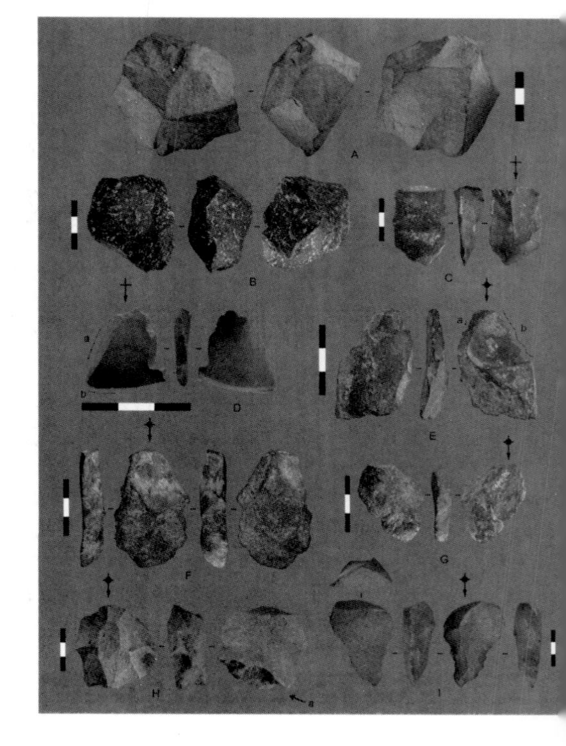

Stone artifacts from Mata Menge associated with the fossilized remains of large-bodied *Stegodon*, Komodo dragons, rats, birds and crocodiles. These date to between 880,000 and 800,000 years ago.
Photo composite: Mark Moore.

A Manggarai whip fight (caci) is a very colorful, exciting
d noisy event enacted to the
companiment of gongs and
uch encouragement from the
dience. Men take turns to try
d cut each other with a buffalo-
de whip, while the other defends
mself with a hide shield.
oto: Douglas Hobbs.

Manggarai welcoming ceremony at Liang Bua
2003. Of the peoples of
ores, only the Manggarai wear
ch sarongs, called *ikat*. *Photo:*
uglas Hobbs.

Wet sieving of deposits excavated from Liang Bua in 2004 by Gaba Gaur and Rius Laru. This system, designed by the Manggarai, probably increased excavated finds tenfold.
Photo: Mike Morwood.

Bert Roberts and the 11,000-year-old white tuffaceous silts in Sector VII of the Liang Bua excavation. The remains of *Stegodon* and *Homo floresiensis* were only subsequently found below these silts. Skeletal and diagnostic behavioral evidence for modern humans was only found above. *Photo: Mike Morwood.*

Thomas Sutikna and Benyamin Tarus excavating the LB1 skeleton at Liang Bua on August 8, 2003. *Photo: Wahyu Saptomo.*

The LB1 skeleton assembled at ARKENAS. In addition to her tiny stature and brain, LB1 has a number of very primitive cranial and postcranial traits, as well as some that are unique, unlike any other hominid species. *Photo: Bill Jungers.* →

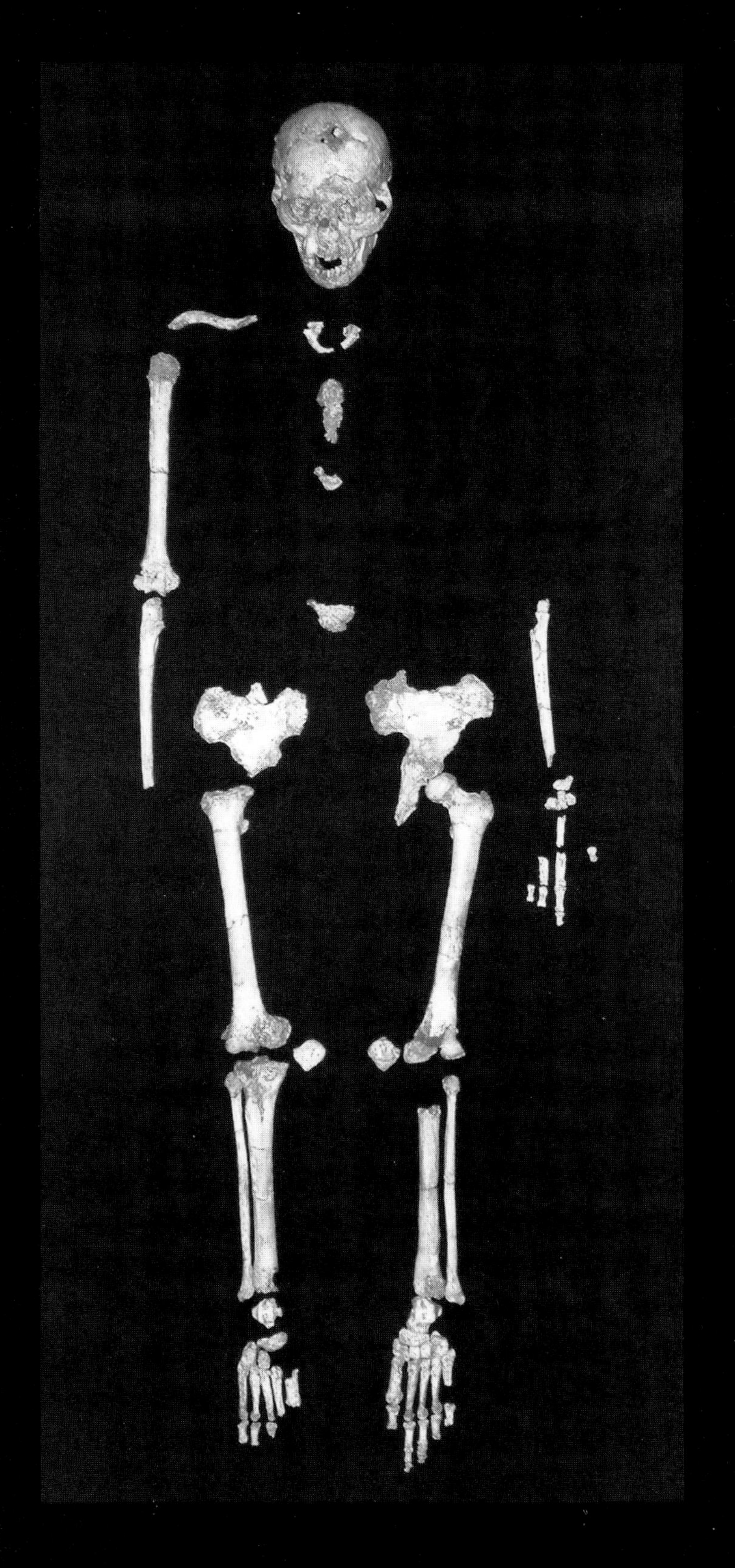

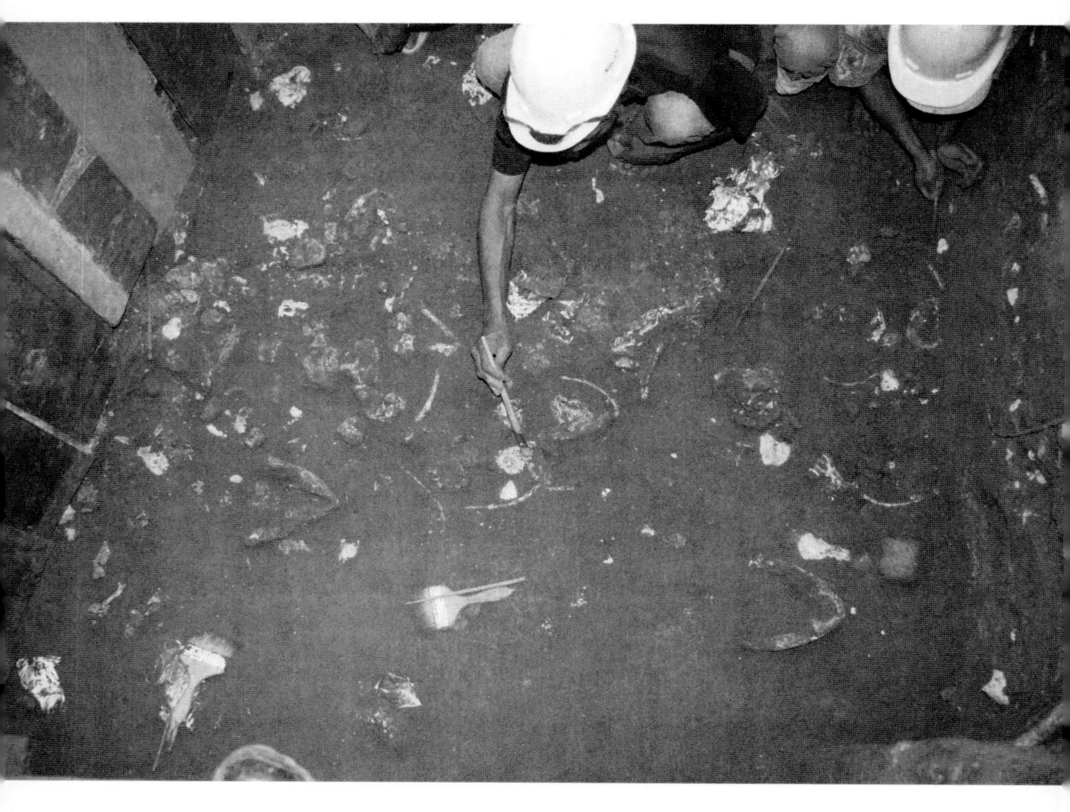

A 15,000-year-old occupation level in Sector XI, Liang Bua, that contained the remains of *Stegodons*, Komodo dragons, rats, bats, birds and *Homo floresiensis*, together with abundant evidence of making and using stone artifacts, and a hearth. *Photo: Mike Morwood.*

Stone artifacts from Liang Bua dating to between 95,000 and 11,000 years ago. The same artifact technology and range of tool types occur at *Stegodon* fossil sites in the Soa Basin, where they are dated to between 880,000 and 650,000 years ago. *Photo composite: Mark Moore.* ➔

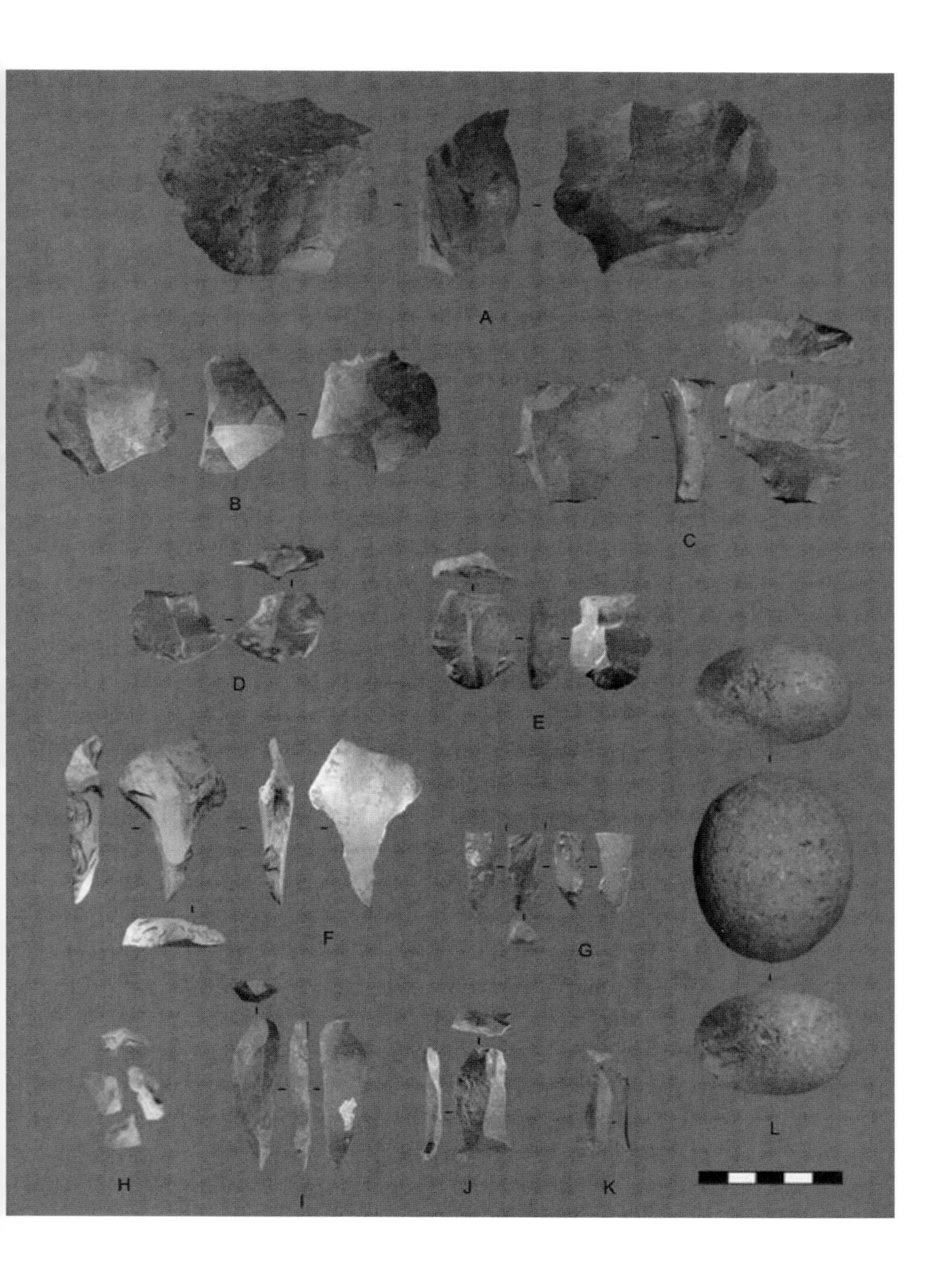
A
B
C
D
E
F
G
H
I
J
K
L

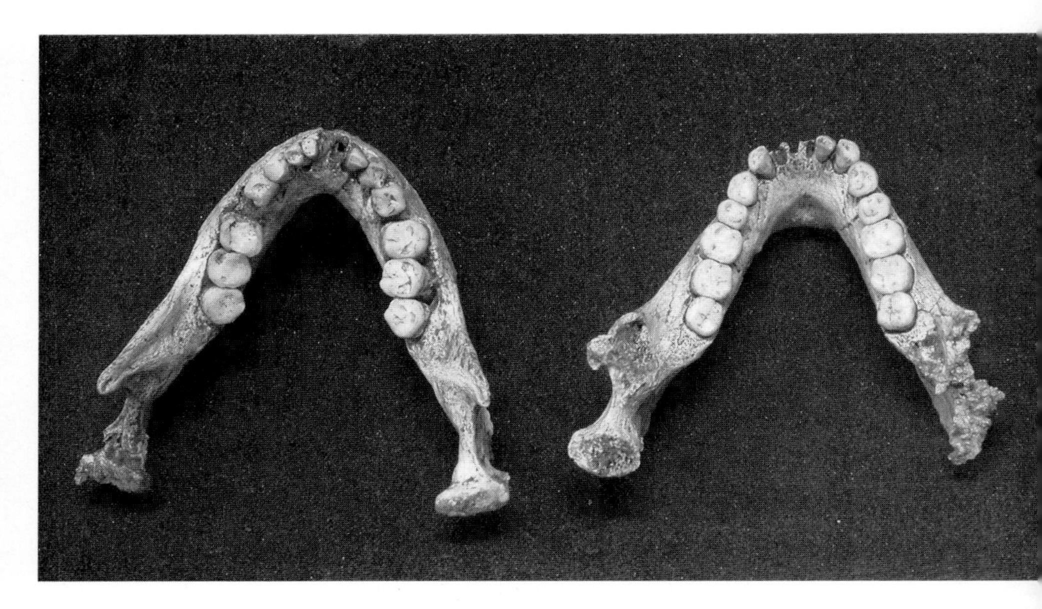

Comparative plan views of the LB1 and LB6 lower jaws. They share a number of traits, including the absence of a chin, but the LB6 jaw was from a smaller individual. *Photo: Thomas Sutikna.*

Tony Djubiantono visiting the Liang Bua excavation in 2004 with Kira Westaway. *Photo: Mike Morwood.*

Nature cover, October 28, 2004.
Courtesy: Nature.

The 2004 raft trip from Sumbawa to Komodo Island that was featured in the National Geographic film by David Hamlin in October 2004. The craft was remarkably stable, but was also unwieldy, slow and swept south by the strong currents. *Photo: Wahyu Saptomo.*

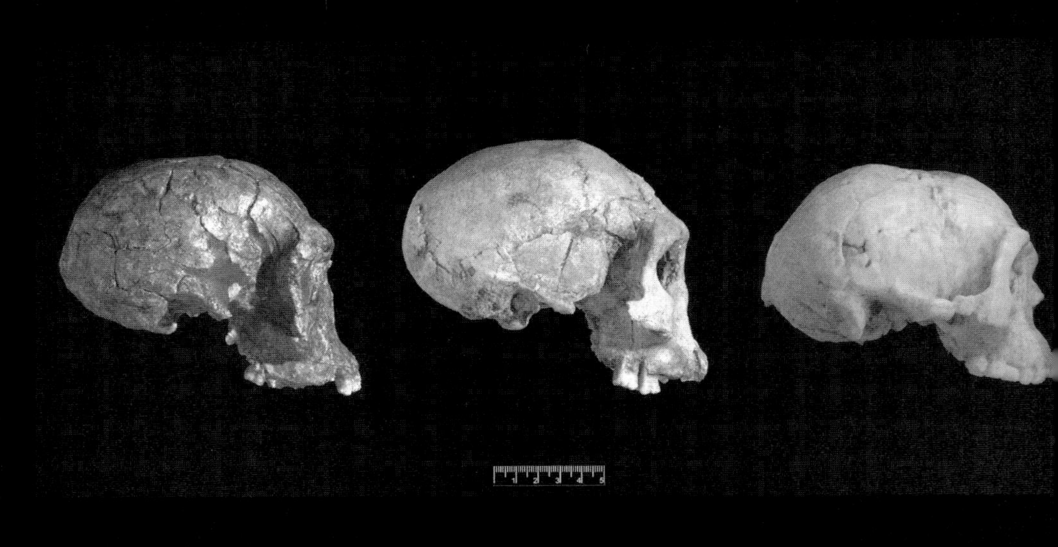

Skulls of 1.9 million-year-old *Homo habilis* from East Africa, 1.8 million-year-old *Homo georgicus* from Dmanisi in Georgia and 18,000-year-old *Homo floresiensis* from Flores. Despite being separated by 2 million years and 9,000 kilometers the *Homo habilis* and *Homo floresiensis* skulls are remarkably similar. The Dmanisi hominids are taller, bigger brained and have modern body proportions. We conclude that *Homo floresiensis* is descended from habilis-like hominids in East Asia. *Photo: Chris Stringer, Natural History Museum, London.*

Upon further analysis, Peter was now saying that LB1 has even more archaic traits than Leakey's two-million-year-old *Homo habilis*.

Modern human woman with Hobbit. Note that there are many traits that distinguish these two hominid species—not just brain size and stature. (DRAWING: PORTIA SLOAN)

Surely, I rejoined, that would simply confirm that the original source population is even earlier on the line of *Homo* development. Peter disagreed.

I could see why, because what this implied was bucking not one, but two, of paleoanthropology's most basic premises: that the genus *Homo* originated in Africa, and that an early type of *Homo erectus* was the first hominid to leave Africa about 1.8 million years ago. My challenge to Peter's designation amounted to suggesting that the earliest good evidence of the first human may not be African at all, but from a lost island world that was, as far as paleontology went, on the forgotten side of the planet.

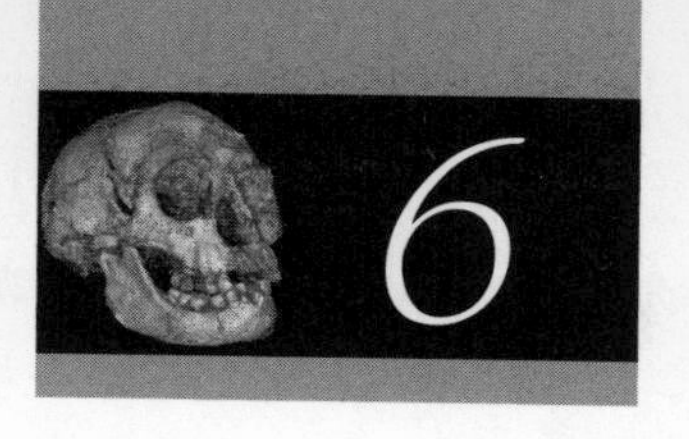

Out of Asia

Our discovery suggested that Asia may have played a much more prominent role in human evolution and for much longer than generally believed. It also indicated how little is really known about early hominid evolution and dispersal anywhere. In fact, so many pieces of the jigsaw are missing that paleoanthropology should be in a continual state of flux as new evidence is unearthed, but historically this is not how many in the discipline have reacted to new, unexpected findings. The hominids found under a medieval castle in Dmanisi, Georgia—deep into the Eurasian landmass, thousands of kilometers from Africa, are a good example of actual evidence disrupting cherished convictions.

Found in the 1990s, and then dated to 1.8 million years old, the Dmanisi hominids are currently the most ancient undisputed hominid fossils outside Africa. At a stroke they almost doubled the accepted date for the first dispersal of humans out of Africa, which was thought to have occurred a little over a million years ago, based on *Homo erectus* finds in Java. The Dmanisi hominids predate the appearance of large-bodied, large-brained, humans. Some are small bodied (1.4 meter stature), small brained (ca. 600 cubic centimeters)

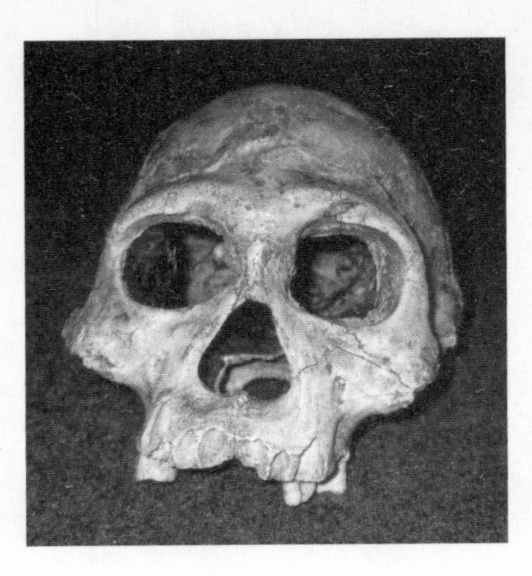

The 1.8 million-year-old Dmanisi hominid skull D2700 shares specific traits with LB1. Some of the Dmanisi hominids are small bodied (about 1.4 meters tall), with small brains (about 600 cubic centimeters).They probably represent a new species, Homo georgicus, *which may have given rise to much later classic* Homo erectus *populations in East Asia. Some of the traits of* Homo floresiensis *are more primitive than those of the Dmanisi hominids, which had the same limb proportions as* Homo ergaster, Homo erectus *and modern humans. The Liang Bua hominids appear to have evolved from a similar, but earlier population.* (PHOTO: DEBBIE ARGUE)

and have primitive characteristics previously only seen in the earliest members of genus *Homo* and australopithecines. They do have modern body proportions. They have been variously described as early Asian *Homo erectus*; earlier *Homo ergaster*; even earlier *Homo habilis*; a new species *Homo georgicus*; and a new subspecies, *Homo erectus georgicus*. This assemblage can definitely be described as taxonomically challenging.

Another problem raised by the Dmanisi hominids is that they are about the same age as the earliest identified species of *Homo* in Africa—and there is certainly no guarantee that the site represents the oldest hominid presence in Asia! Any older findings, and there will be even more reason to question the present disciplinary preoccupation with Africa as the one and only source of all stages of hominid evolution.

And then there is LB1. On Flores, an oceanic island east of mainland Asia and always a difficult colonization prospect, we have found a hominid, LB1, who is smaller in stature, with a smaller brain and more primitive characteristics than any of the Dmanisi specimens. Her more apelike body proportions, for instance, are similar to *Homo habilis*, *Australopithecus afarensis* and particularly *Australopithecus garhi*, but differ from other hominid species for which proportions can be calculated. Endemic dwarfing goes only part of the way to explain her diminutive brain and body size. The ancestral population for *Homo floresiensis* almost certainly underwent size reduction on the island, but general biological principles indicate that proto-hobbits cannot have been a large-bodied hominid species such as *Homo ergaster* or *Homo erectus*—a conclusion supported by the range of primitive traits retained by LB1 and her kin.

With the principle of biogeography that Alfred Russel Wallace first established in his Sarawak Law—that every species has come into existence in about the same place and time as its nearest relative—then, with the current evidence, you would be bound to say that the assumption that early hominids always dispersed out of Africa and across Asia is unwarranted. You could say with equal justification that *Homo* evolved in Asia and moved to Africa. It's actually not such an original idea, as seen in the history of hominid research, where the main focus has swung from Africa to Asia and back to Africa. Maybe now, with new evidence and renewed interest in Asia, a more balanced perspective will emerge.

Darwin saw that comparative anatomy and embryology demonstrated that apes and humans had descended from some common ancestor, and suggested in his *The Descent of Man*, published in 1871, that humans had originated in Africa because African apes were anatomically our closest relatives:

> [I]t is somewhat more probable that our early progenitors lived on the African continent than elsewhere. But it is useless to speculate on this subject . . . since so remote a period has certainly undergone many great revolutions, and there has been ample time for migration on the largest scale.

But until Eugène Dubois, the young professor of anatomy whom we met earlier, went on a quest for hard evidence in the fossil record, there was only speculation on what human ancestors would have looked like.

Born in 1858, the year before Darwin published his *On the Origin of Species*, Dubois grew up under the influence of the theory of evolution. By the time he had become a young medical doctor, in 1884, the theory was widely held, except when it came to humans. Not that this seemed to bother Dubois. By all accounts, he was a contrary person, and it was perhaps his prickly nature that forced him to go in search of direct proof of what few thought could be found—the tie between modern humans and the animal world—to pursue the bold yet utterly original idea of finding the so-called missing link.

Dubois' great insight was to realize that although anatomy and embryology indicate evolution's working, only fossils could demonstrate what had really been. Furthermore, in his quest to find this fossil evidence for human evolution, he accepted an alternate

Eugène Dubois, who went in search of fossil evidence for human evolution and, in 1891, found his "missing link" between apes and humans in Java.

argument, first put forward by the German biologist Ernst Haeckel, that the behavior and anatomy of gibbons and orangutans in tropical Asia indicated a closer relationship with us: gibbons, for instance, are apes that live in faithful couples, sit upright on branches with their spouses, and on the ground even walk upright in a fashion, with long arms held out. It was clear to Dubois that the search for the ancestors of "man" lay in the Orient. The fact that the East Indies was a Dutch colony also made it an easier choice.

In 1887 Dubois became a medical surgeon in the Royal Dutch East Indies Army and boarded a ship bound for Sumatra. Despite winning official support, which enabled him to become a full-time fossil hunter from early 1889, his endeavors had only limited success in Sumatra, and in 1890 he transferred to East Java, where fossil finds, including the Wajak skull, had already been made during geological surveys. He also transferred the focus of his search from caves to open-air fossil exposures along the Solo River. Fossil finds made by his two military overseers and a large crew of forced convict laborers included the remains of *Stegodon,* elephant, hippos, rhinos, bovids, pigs, hyenas and cats, as well as a fragment of hominid lower jaw, with a canine and two premolar sockets, from the site of Kedung Brebus. On the basis of similarities with fossils from mainland Asia, including India, he estimated that the Java finds were likely to be of Early Pleistocene age—the last glacial period, which we now know dates from 1.8 million years to 700,000 years ago.

Dubois then began excavations of a fossil-rich layer of sediment on the banks of the Solo River at Trinil. In less than a month, he began to hit hominid pay dirt: a tooth of a primate emerged from a layer of volcanic sediment. It was a molar, the third molar of the upper right jaw. Erring on the side of caution, for the time being, Dubois designated the find as a "chimpanzee." But the following October, a peculiar skullcap was unearthed. It had a low forehead and jutting eyebrow ridges, like an ape, but a high vault like a human, and a cranial capacity of 1,000 cubic centimeters. From the skullcap it was clear that his "chimpanzee" was no ordinary ape, being substantially more humanlike. Dubois knew he had discovered something important. But he needed something else, some other evidence. Dubois' prize came the following August 1892, when an almost complete left thighbone was uncovered, in the same-aged layer as the molar and skullcap, but fifteen meters upstream. Amazingly, it was indistinguishable from a modern human's.

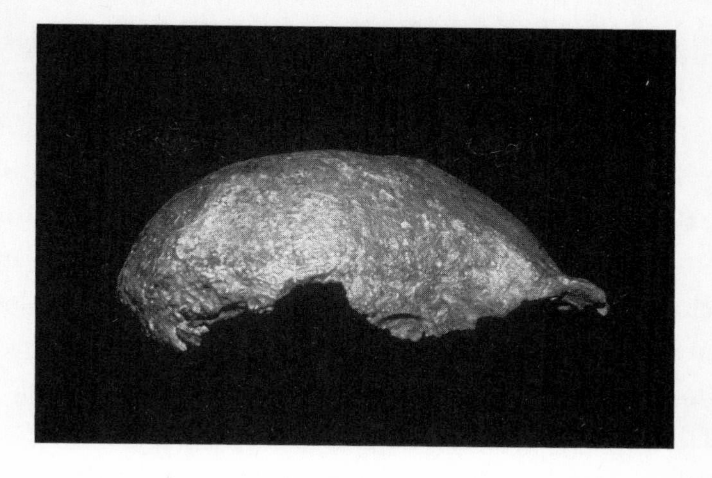

Left side of the Pithecanthropus erectus *skullcap found by Eugène Dubois in 1891 at Trinil, Java. This is the type specimen for the species later designated* Homo erectus. (PHOTO: NATURALIS, LEIDEN)

Dubois was sure that the molar, skullcap and thighbone belonged to one and the same individual, who was in no way equipped to climb trees. On the contrary, it was obvious from the construction of the femur that the bone fulfilled the same mechanical role as in the modern human body. It was clear that the ape-man of Java stood upright and moved like a human. Dubois named the find *Pithecanthropus erectus*, meaning "upright ape-man," saying it was a transitional form—a missing link—and man's ancestor.

By 1900, six years after Dubois had published a monograph on his find, nearly 80 articles and books had been published in response, not including popular articles of which there were countless numbers. Thus, for the first century of paleoanthropology, starting with the discovery of the first Neanderthal in 1850, Eurasia was considered central, not peripheral, to the evolution and dispersal of primates, including humans. In ideas that encapsulate the thinking of the early part of the 20th century, well-respected mammalian paleontologist and biogeographer William D. Matthew, in 1915, wrote in a paper titled "Climate and Evolution" that "most authorities are today agreed in placing the centre of dispersal of the human race in Asia," pointing out the great plateau of Central Asia as the likely area. Matthew's argument was that Asia was of paramount importance in the evolution and dispersal of primates. He emphasized that immediately around the borders of the great plateau are regions of the earliest recorded civilizations, in Egypt, India and China. Nor did he think much of the idea that humans were primarily adapted to tropical climate, and with the considerable bias of the time claimed that

> . . . it will not be questioned that the higher races of man are adapted to a cool-temperate climate, and to an environment rather of open grassy plains than of dense moist forests. In such cool climates they reach their highest physical, mental and social attainments.

Although rife with insulting racial and political prejudices, such writing still had a powerful effect on anthropology; the idea that Central Asia was the center of human evolution was ultimately what drove Davidson Black to China, closer to what was thought to be the likely site of human origins. In 1919, Black accepted a medical position at Peking (now Beijing) in north China. In 1929, his excavations at the nearby site of Chou Kou Tien (Dragon Bone Hill) yielded the skullcap of a hominid that Black named *Sinanthropus pekinensis*—Peking Man—who Black conjectured was related to *Pithecanthropus*, a claim vehemently denied by an irascible, now elderly Eugène Dubois, for whom there could only be one missing link: his.

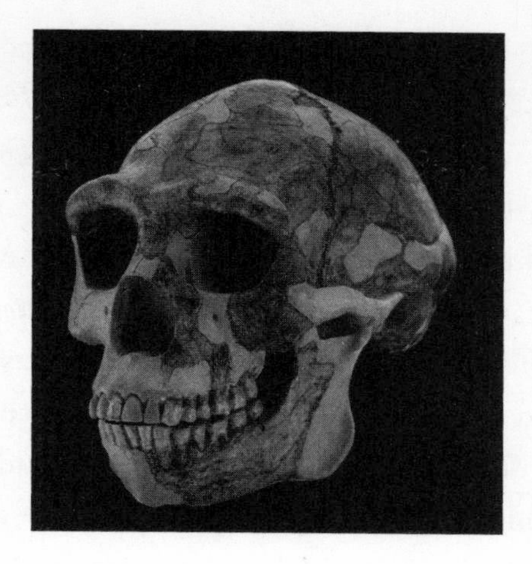

Peking Man reconstruction based on Weidenreich's original cast Homo. (PHOTO: DAVID BRILL, COURTESY OF AMERICAN MUSEUM OF NATURAL HISTORY)

The discovery of other early hominid sites kept East Asia at the leading edge of world paleoanthropology and associated issues. In 1931, Cornelius ter Haar of the Geological Survey of the Dutch East Indies (now GRDC) discovered a fossil deposit at Ngandong on the Solo River, where it cuts through the Kendeng Hills in East Java. His acting director, W.F.F. Oppenoorth, then sent a team to the site to undertake excavations, which lasted until November 1933 and yielded thousands of animal fossils, as well as 11 skull-caps of a new hominid species subsequently named *Homo soloensis*.

Despite the obvious similarities between *Pithecanthropus erectus*, *Sinanthropus pekinensis* and *Homo soloensis*, Dubois maintained that *Pithecanthropus* alone occupied the "missing link" position—that *Sinanthropus* was an early human and that *Homo soloensis* was essentially a modern human similar to Australian Aboriginals. This led to a major falling-out with the newly appointed vertebrate paleontologist at the Geological Survey, and rising star in the field of paleoanthropology, G.H. Ralph von Koenigswald.

Von Koenigswald made extensive surveys and collections at Ngandong, and initiated work at Sangiran, which has since yielded about 60 percent of the world's known *Homo erectus* fossils. In 1936, when one of his assistants, Andojo, found a fossilized child's skull at Mojokerto, von Koenigswald promptly interpreted it as *Pithecanthropus*—an interpretation vehemently denied by Dubois, who argued that the skull was of a "real human child" and had probably been found on the surface and later mixed up with excavated finds. Von Koenigswald backed down for a while, and referred to the new specimen as *Homo modjokertensis*. However, the bitter debate was renewed when further hominid finds were made at Sangiran and also described by von Koenigswald as *Pithecanthropus*, denied in every instance by Dubois, who even accused von Koenigswald of doctoring skulls to strengthen his argument.

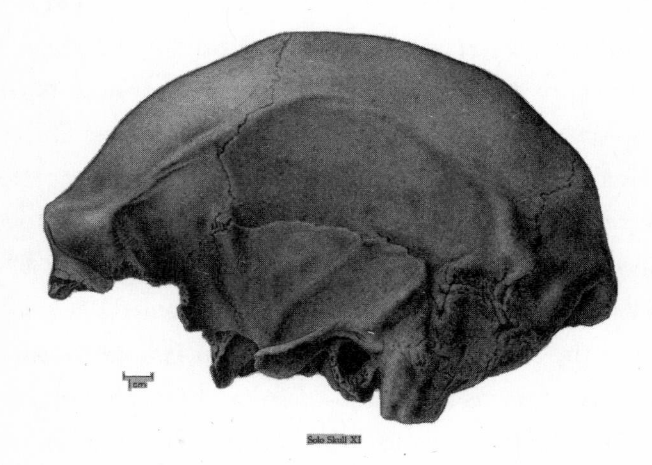

Ngandong Skull XI. During 1931–33, a total of 11 such late Homo erectus *skulls were excavated from the Ngandong Terrace on the Solo River in East Java. Despite obvious similarities with the older* Pithecanthropus erectus *cranium found by Dubois, he dismissed these finds as being modern humans that were similar to Australian Aborigines.* (CREDIT: MIKE MORWOOD, AFTER WEIDENREICH, 1951)

Both men also tried to enlist support from fellow scientists of note. In 1938, Dubois invited Franz Weidenreich, who, after the premature death of Davidson Black, was in charge of the Chou Kou Tien excavations and finds, to visit, with the idea of convincing him that *Sinanthropus* and *Pithecanthropus* were fundamentally different. However, von Koenigswald and his new fossil finds proved more persuasive and he was invited by Weidenreich to visit Peking to see the *Sinanthropus* material. Weidenreich also asked him to bring the new Sangiran hominid fossils to make comparisons. Von Koenigswald accepted the invitation and later wrote:

> Every detail of the originals was compared: in every respect they showed a considerable degree of correspondence. The two fossil men are undoubtedly closely allied, and Davidson Black's original conjecture that *Sinanthropus* and *Pithecanthropus* are related forms—against which Dubois threw the whole weight of his authority—was fully confirmed by our detailed comparison.

Von Koenigswald and Weidenreich published this conclusion in an article in *Nature*, but World War II then interrupted further research and debate, during which time the Sangiran and Mojokerto fossils were safely concealed, while all the original *Sinanthropus* finds were lost. Dubois' last words on the subject, two weeks before his death at age 82 on December 16, 1940, were:

> It is most regrettable, that for the interpretation of the important discoveries of human fossils in China and Java, WEIDENREICH, VON KOENIGSWALD and WEINART were thus guided by preconceived opinions, and consequently did not

> contribute to (on the contrary they impeded) the advance of knowledge of man's place in nature, what is commonly called human phylogenetic evolution. Real advance appears to depend on obtaining material data in an unbiased way.

The demise of Dubois did not end the war of words. In his introduction to the monograph on the Ngandong fossils crania in 1951, Weidenreich replied:

> The limit of tolerance for these human foibles is attained when the proponent of a questionable scientific doctrine endeavours to maintain it against all possible odds by misrepresentation, misinformation, and the suppression of contradictory data, and by insinuating unfairness in opponents of his views. I regret to find it necessary to accuse Dubois of such an attitude.

By that time, the major issue for East Asian hominids, their taxonomy, had been tackled by Harvard evolutionary biologist Ernst Mayr, who lumped a number of specimens together, including the fossils from Chou Kou Tien and Java, into one species—*Homo erectus*. He also distinguished a number of geographical and chronological subspecies, including *Homo erectus pekinensis* from China; and Lower Pleistocene *Homo erectus robustus*, Middle Pleistocene *Homo erectus erectus* and Late Pleistocene *Homo erectus soloensis* from Java. The latter, from the sites of Ngandong, Ngawi and Sambungmacan along the Solo River, are a population with more modern traits. Mayr's taxonomic system has been widely accepted.

A recent study by the Japanese paleoanthropologist Kaifu and his colleagues of all hominid teeth and lower jaws recovered from Sangiran in Central Java, confirms that substantial changes took place in the local *Homo erectus* population over time. The study shows that two strikingly different hominid populations occupied the site separated by a gap of 400,000 to 600,000 years. The Lower Pleistocene *Homo erectus* population about 1.2 million years ago had some very primitive traits harking back to the earliest members of *Homo* in Africa. In contrast, those present at Sangiran during the Middle Pleistocene, from around 800,000 years ago, had traits similar to those of Peking Man, thus banishing the old idea that *Homo erectus* was some static stuck-in-the-mud hominid. Did one species evolve into the other, or was Java invaded by different *Homo erectus* populations in a world of mobile hominids? Like other mammals in Southeast Asia, *Homo erectus* seems to have had an evolutionary history as dynamic and complex as the landscape.

Since World War II, many more *Homo erectus* fossils have been found at Sangiran and along the Solo River in Java, usually by local people going about their daily activities, in canal construction, diving for black sands or digging wells. By the 1960s the sheer number of finds seemed to indicate that hominids had emerged in East Asia and then spread westward to Africa and Europe.

Nevertheless, while Piltdown Man was still in vogue, while Eugène Dubois was still smarting from criticism of his *Pithecanthropus erectus*, and before Davidson Black had excitedly described his first *Sinanthropus* tooth, going out on a limb to state that it was a relic of a new type of early man, a true ancestor of modern humans had already been discovered, not in East Asia but in Africa.

Dubois to the contrary, the link between apes and humans is found among the australopithecines. Arthur Dart, an Australian anatomist working in the newly founded University of Witwatersrand in Johannesburg, South Africa, found the first australopithecine in 1924. Dart discovered the fossilized skullcap and lower jaw of a very young australopithecine child, in two-to three-million-year-old lime debris left by a mine excavation into the remains of a cave near Taung. He had heard that there were fossils in the lime of the quarry, and had asked the quarry owner to send him any fossiliferous rock that turned up. In due course large boxes of limestone arrived at Dart's house. Fossicking about in the boxes, he found a cast of the interior of the skull. Looking at the convolutions and furrows, he thought it might have been a cast of a baboon, but it was too big to fit inside a baboon's skull. Scrabbling around inside the box he found a piece of skull still encased in rock. After 73 days of chipping and picking with a small chisel, and then his wife's knitting needles, he was astounded to reveal a skull of what he ascertained was a six-year-old child. The skull was too high and rounded, and the face too small, for it to be a baboon or chimpanzee. Turning the skull over, he noticed that the foramen magnum—the hole where the spine ascends to the brain—was at the bottom of the skull. The child had walked upright!

An ape in South Africa, let alone one that walked upright, seemed impossible, yet the evidence was staring him in the face. He stared back and was seized by the notion that it was some kind of missing link. He dashed a paper off to *Nature* describing the find and naming it *Australopithecus africanus*, meaning "southern ape of Africa." It was quickly nicknamed "Taung Child," and found to have had a number of outstanding human features that Charles Darwin had predicted for early hominids: reduced canines compared to the other great apes, bipedalism, and changes in the structure of the brain. Further confirming the closeness of the species' relationship to *Homo sapiens* were the small, flat-wearing canines, and the shape and pattern of the brain, which was imprinted on the inside of the skull. Although small, the child's brain was

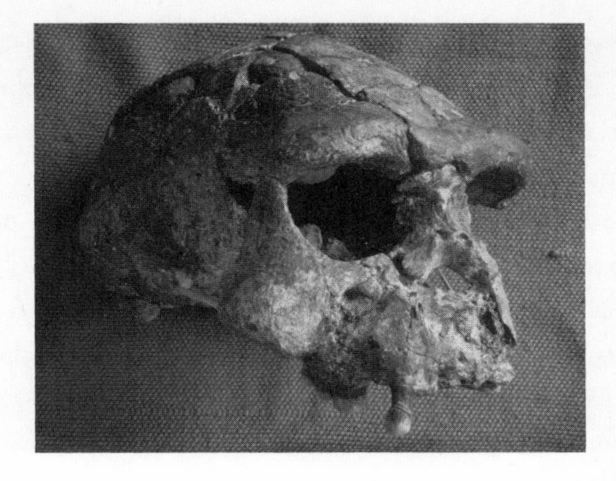

Homo erectus *in Java. Sangiran 17, the most complete* Homo erectus *skull found in Java and around 750,000 years old. This shares many characteristics with* Homo erectus *of about the same age or less in China. However, older* Homo erectus *remains from Sangiran have a number of features otherwise found only in early African members of genus* Homo. (PHOTO: FACHROEL AZIZ)

slightly larger and structurally different from that of a comparably aged ape.

The British science establishment, however, was not impressed. Sir Arthur Keith, at the center of the paleoanthropological universe, and enamored of Piltdown Man, dismissed Dart's find out of hand, saying that, at best, it was a new genus of chimpanzee or gorilla. Before Dart could make his case, in the tradition of absentminded scientists, he lost the Taung Child skull. In 1931, he was invited to an anthropological congress at which the star attraction was Peking Man, but mistakenly he left the Taung Child in the back seat of a London cab. Later that night, the taxi driver finally noticed the package and turned it over to the police, who thought they had a murder on their hands. Meanwhile, Dart himself had notified the police when he realized the Taung Child was missing, and picked it up before returning to South Africa. It took 17 more years and several more discoveries of australopithecines by different scientists before the Taung Child, and australopithecines in general, were finally recognized as hominid.

But it would take major discoveries by the most paradigm breaking of all paleontologists, Louis Leakey, with his *National Geographic* personality, to wrench the attention away from Asia and focus it on Africa's vast Rift Valley—which stretches 3,000 kilometers down East Africa from Ethiopia south to Malawi, exposing layers of rock and fossils dating as far back as six million years. Louis and his wife, Mary Leakey, spent years in the shadow of World Heritage Ngorongoro Crater in Tanzania, in a desolate gorge with the hollow, ancient-sounding name of Olduvai. There, in 1930, Louis found what he recognized to be deliberately shaped stone tools. To the untrained eye they were merely hundreds and thousands of broken cobbles. But Louis saw them for what they were—primitive stone tools, with a flake or two knapped off one end to form a crude but effective cutting edge. He became obsessed with the identity of the knappers: human for sure because they made tools, but awfully primitive. For Louis, solving the mystery became an obsession.

For 30 years he and Mary used every scrap of money they had searching for the maker of those tools. Then, one day in August 1959, Mary, working alone while Louis lay in a malarial fever, spotted some massive teeth and a shattered skull eroding out of a gentle hill slope. She bolted to the camp and with a feverish Louis in tow raced back to the site. When the pieces had been put together it turned out to be one of the most magnificent hominid skulls ever found, with massive cheekbones and a keel running down the center of the skull, too primitive to be human, and the wrong sort of skull to be associated with the tools. It was assumed that something so primitive could not have made the tools scattered at the bottom of Olduvai Gorge. It was more like an australopithecine, but because the teeth were much larger, Louis placed it in a new genus, *Zinjanthropus boisei*; *Zinj* deriving from an Arabic word meaning East Africa, and *boisei* to honor a financial backer, Charles Boise. "Zinj," as the skull was nicknamed, became an instant sensation.

At the time of this discovery, the last common ancestor for all primates—not just the human lineage—was thought to have lived two million years ago. The human stem, beginning with australopithecines, which had only just been accepted as hominids, was thought to have split from the other great apes some 900,000 years ago. Change was on the way, however. In 1960, Italian scientists trying to date volcanic deposits around Rome developed a revolutionary dating technique that relied on the rate at which a radioactive potassium isotope decayed into a stable argon one to find how much time had elapsed since a volcanic rock solidified. Zinj was closely associated with a layer of volcanic tuff at the bottom of Olduvai Gorge that could be dated using the new technique, and he became the first early hominid in the world to have a reliable date. And what a date it was. At a time when many thought the Pleistocene—Zinj's time—began about 200,000 years ago, Zinj turned out to be a whopping 1.8 million years old.

Perhaps surprisingly, considering that Zinj made Louis Leakey famous, and archaeology sexy, it was a disappointment for Leakey, who insisted that, while interesting, Zinj was not a human. He believed that a straight line went back from *Homo sapiens* to an always cruder but always recognizable sapient forebear. He badly wanted a *Homo,* and this find was not one.

In 1964, his dream came true when he rocked the world yet again, with the announcement of another hominid fossil at Olduvai, this time not of an australopithecine but of a true human, which he named *Homo habilis*—"Handy Man," a name suggested by Raymond Dart—in honor of the species having been the true maker of associated stone tools. The real surprise was the age of this new *Homo*. It was the same age as Zinj.

In one strike the Leakeys had dismissed another of the dogmas, that hominids evolved one species at a time. The logic for this dogma was that there was simply no ecological space on the planet for more than one unique culture-bearing species at a time; there was only *Homo sapiens* today, and there was only *Homo erectus* yesterday. The implicit scenario was of a slow, single-minded, almost preordained transformation of the hunched ancestral hominid into the glorious modern *Homo sapiens*. It was powerfully seductive. But instead of this, the tree of man began to look like a bush, and the tangle of species being discovered was proving confusing. Partly this reflects the nature of the early hominid fossil record, which is sparse, intermittent and uneven. But while Leakey continued to stun the world, it was the *Australopithecus afarensis* skeleton nicknamed Lucy who riveted everyone's attention to Africa, leaving the prehistoric mysteries of Asia blowing in the wind.

By the time Lucy's discoverer, paleoanthropologist Donald Johanson, came on the scene in 1973, at least four different types of African hominids had been discovered—two australopithecines and two types of *Homo*. There was some considerable confusion about who had given rise to whom, and Johanson joined the frenzy of activity, deciding that he would have a shot at sorting out the mess by finding a hominid in a new site he'd recently canvassed, the Afar region in Ethiopia. Johanson was lucky. During his first field season, he turned up a hominid knee joint in sediment at least three million years old—as old or older than any other hominid then known. It doesn't sound like much but, for the times, it was a pivotal find. In 1973, no one had imagined that such early, primitive australopithecines could be fully bipedal—like us. But the modern-looking knee joint said they were.

The next year, Lucy turned up. She was "the oldest, most complete, best-preserved skeleton of an erect-walking human ancestor that has ever been found," ostensibly the grandmother of us all. At the time, australopithecines were considered the earliest hominids and represent the longest-lasting era—about four million years—of pre-stone tool-making hominid forms, and it is among them that our ancestors will be found. With a receding forehead and a brain volume not much larger than a chimpanzee's, Lucy was not particularly smart—there is no definite evidence that australopithecines made stone tools. Nevertheless, her brain was undergoing a major neural reorganization that involved the expansion of those areas associated with cognition, categorization, symbolization, and speech.

Between three and four million years ago, *Australopithecus afarensis* ranged throughout the woodland and savanna belt from the Atlantic Ocean across the Sahel to the Cape of Good Hope. They enjoyed the lakes, gallery forest, wooded savanna and open grassy patches together with giraffes, cattle, pigs, hippopotamus, rhinoceros, early horses and elephants, which had also recently evolved, or migrated from the north into Africa, to take advantage of the new, more open environments in the run-up to the Pleistocene Ice Age, when the major deserts, and ice, clenching the northern lands, would form formidable barriers. Under the same pressures as other animal lineages that rapidly speciated during the changing environments from 2.6 million years ago, at least two separate hominid lineages emerged from *Australopithecus afarensis*—an australopithecine lineage and one that led to the *Homo* line.

The predominant view now, first argued by Johanson and Tim

White of the University of California–Berkeley, is that *Australopithecus afarensis* was the direct ancestor of all later hominids, including genus *Homo*, as well as later australopithecines, such as the robust *Paranthropus aethiopicus* found in East Africa and the gracile *Australopithecus africanus* from South Africa. The tiny, partial skeleton of Lucy, which is the best known *Australopithecus afarensis* find, does have a superficially humanlike body.

Another view, promulgated by Lee Berger, director of the paleoanthropology unit at the University of Witwatersrand in Johannesburg, holds that *Australopithecus africanus,* which has relatively longer arms and shorter legs compared to Lucy, represents the more primitive, arboreal hominid condition, and cannot therefore be derived from *Australopithecus afarensis*. Instead, he concludes that the two were contemporaries, and that *Australopithecus africanus* is the most logical ancestor for genus *Homo*.

Whatever is the case, the earliest members of genus *Homo* appeared in Africa by 2.3 million years ago. The best-dated evidence comprises a single upper jaw from the Hadar region of Ethiopia. This jaw has the broad, parabola-shaped dental arch that is diagnostic of our genus, but the evidence is insufficient to allocate it to a specific species. Stone artifacts first appear in the archaeological record about this time. These earliest knappers were competent and had a good understanding of the properties of suitable stone, which they would carry up to several kilometers from its source. No modern apes, including chimps and orangutans, are capable of such complex behavior or planning. The makers of stone tools are typically referred to as members of genus *Homo,* although there are reports of animal bones showing signs of having been cut associated with the 2.5-million-year-old remains of *Australopithecus garhi* at Bouri in Ethiopia—no conclusion is safe from new evidence.

By two million years ago, at least two species of *Homo* can be identified in East Africa—*Homo habilis* and *Homo rudolfensis*. Both retain a number of primitive australopithecine traits, but *Homo habilis* is smaller in body and brain size, with relatively long arms and short legs. However, the fossils attributed to the species are so diverse that they could represent more than one type of hominid. Some have large, flat faces, while others have human-like faces but small brains. Others again have a crest on the top of the skull similar to one of the robust-looking australopithecines that were around at the time. So it's not surprising that some paleoanthropologists suggest that *Homo habilis* simply represents a grab bag of assorted hominid fossils from around two million years ago. This is quite possible considering that *Homo habilis* shared its era with several other hominids, including *Paranthropus robustus*, *P. boisei, H. rudolfensis* and, later, *H. ergaster.*

Homo rudolphensis *and* Homo habilis*—the two earliest identified species in genus* Homo. *The latter species has a small brain size of about 500 cubic centimeters, is about a meter tall, and has the primitive "ape-like" limb proportions with long arms relative to legs, as does LB1.* (FROM STRINGER AND ANDREWS, 2005)

Homo ergaster (Working Man) arrived on the African stage quite suddenly about 1.75 million years ago, but from which earlier *Homo* lineage, or where, remains uncertain. Again there is little actual evidence for this evolutionary breakthrough. The Nariokotome *Homo ergaster* skeleton is the best-known representative of this species that is characterized by a radical departure from previous forms of *Homo*. The new hominid on the block was tall—at 175 centimeters, twice the size of the local australopithecines—and rangy, with long legs and modern African body proportions. Adult males of the species were only 10 to 20 percent heavier than females, again the same as for modern humans, and quite unlike australopithecines, whose males could be twice the size of females. Overall *Homo ergaster* was so unlike anything previously found that the species seemed to appear as if by magic.

Perhaps because of its tall, striking, "human" aspect, most of today's paleoanthropologists think that *Homo ergaster* was the first hominid capable of migrating out of Africa, striding proudly, carnivorously, and cleverly (at 900 cubic centimeters, it had a large brain) across the savannas of the world, armed with stone tools—the "Out of Africa 1" scenario. *Homo ergaster* is also assumed to have been an African forerunner of the much later East Asian species *Homo erectus*, and a direct ancestor of modern humans. However, the general principles of evolutionary ecology and recent discoveries now seriously challenge some of this anthropological orthodoxy.

The evolution and dispersal of hominids occurred at an extraordinary time in our planet's history; it was a time when two of the world's largest rivers (the Yangtze and the Yellow River) had only recently come into existence, streaming off the newly risen Tibetan Plateau, now the roof of the world. A gargantuan feature—easily

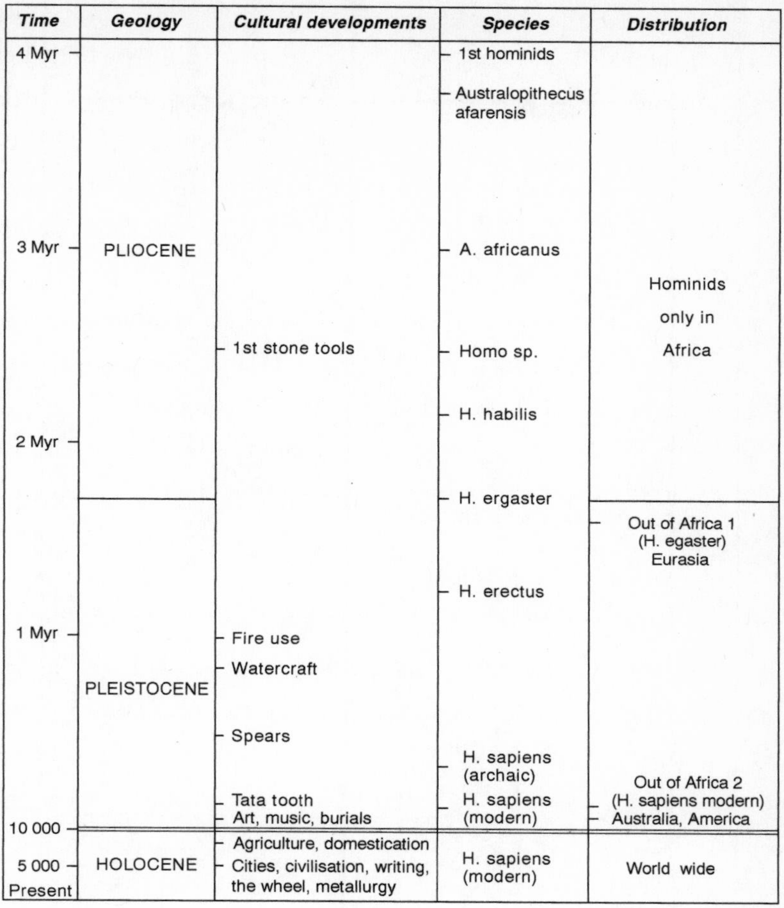

Time	Geology	Cultural developments	Species	Distribution
4 Myr			1st hominids	
			Australopithecus afarensis	
3 Myr	PLIOCENE		A. africanus	
				Hominids only in Africa
		1st stone tools	Homo sp.	
			H. habilis	
2 Myr				
			H. ergaster	
				Out of Africa 1 (H. egaster) Eurasia
			H. erectus	
1 Myr		Fire use		
	PLEISTOCENE	Watercraft		
		Spears		
			H. sapiens (archaic)	
		Tata tooth	H. sapiens (modern)	Out of Africa 2 (H. sapiens modern) Australia, America
10 000		Art, music, burials		
		Agriculture, domestication		
5 000	HOLOCENE	Cities, civilisation, writing, the wheel, metallurgy	H. sapiens (modern)	World wide
Present				

Hominid timeline for the last four million years that includes the conservative scenarios for "Out of Africa 1" for Homo ergaster *around 1.8 million years ago and "Out of Africa 2" for modern humans around 100,000 years ago.* (CREDIT: MIKE MORWOOD)

the size of Western Europe—the Tibetan Plateau emerged with the collision of India into China. It was a profound reorganization, gradual for the first 20 million years; then, on an earthly scale of things, virtually thrusting skyward recently, in the last few million years, to altitudes of more than 5,000 meters. Global in its impact,

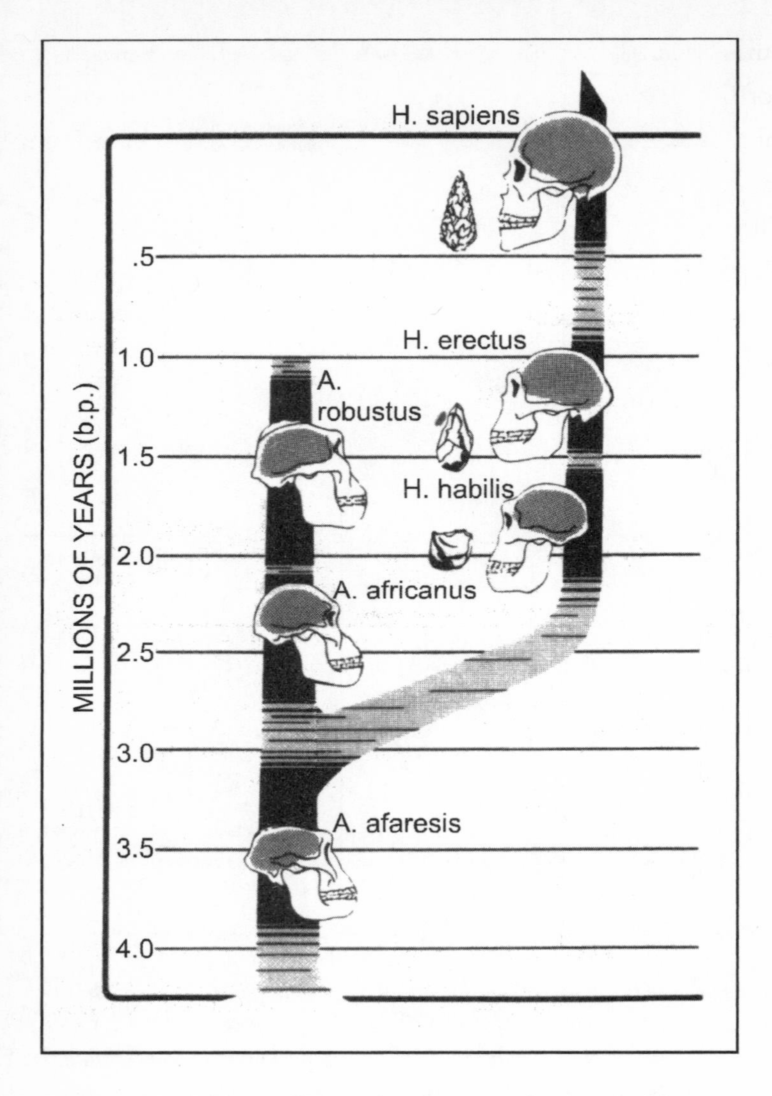

The conventional view of hominid evolution, as first proposed by Don Johanson and Tim White. (CREDIT: MIKE MORWOOD, AFTER JOHANSON AND EDEY, 1981)

the tremendous increase in the elevation of such a large slab of the planet changed even the swirling pattern of the atmosphere, blocking the summer monsoonal air sweeping into India, and eventually causing desert conditions in northwestern China. It also pushed the

Eurasian high pressure system, which had squatted over China, north over Siberia, from where strengthened winds howled back into Eurasia. Farther afield, New Guinea, with Australia at its back, had pushed well into the tropics, blocking the Indonesian seaway. This path of warm South Pacific water was replaced by colder North Pacific water. These and other changes tipped the planet into a cooling phase, causing grassland and savanna to spread from West Africa to north China.

At the time, neither the deserts of the Sahara nor those of Southwest Asia existed, and the regions still supported trees. From a migrating animal's perspective Africa and Asia, with their similar environments, were all fair game. Near the end of the Pliocene, a time of global cooling and drying that began about five million years ago, and the onset of Pleistocene glacial conditions around two million years ago, the same types of animals existed in the Levant as in eastern Africa. These included giraffes, cattle, antelope, elephants, and big cats: the first lions, leopards and spotted hyenas that made Europe their home were enormous, a feature suggesting good times and an abundance of large game. Global environmental changes also forced the pace of evolution. For example, 2.6 million years ago, when the Arctic ice cap and the modern form of Antarctica were established, a wave of extinctions occurred in well-documented groups, such as African antelope, and forest-adapted species were replaced by others better suited to savanna. Although not similarly well documented, a burst of evolutionary activity probably characterized the hominid sequence around the same time—including emergence of genus *Homo.*

The history of mammals from about four million to two million years ago is generally one of speciation, as well as infrequent but widespread dispersals into and out of Africa: at least three events are so far known at 1.9, 2.6 and 3.7 million years. Horses, for instance, which had evolved in North America, trotted over the Bering Strait about three million years ago, and into Europe around 2.6 million years ago, before finally arriving in Africa two million years ago, where they diversified into modern zebras. Elephants, which evolved in Africa, did the exact reverse: they dispersed out of Africa to reach Europe between three and two million years ago and North America by two million. Even the great cats that we think of as African are not, having evolved from an Asian ancestor. Their 11-million-year evolutionary lineage involved at least 10 intercontinental migrations. The immediate ancestor of the quintessential African lion spread into Africa around the era of *Australopithecus afarensis*, perhaps 3.7 million years ago. The cheetah, too, originating in North America from a puma-type ancestor, dispersed into Africa at the same time, filtering through the Pliocene savanna landscapes, not recognizing the fiction of a bounded Africa.

Did hominids move out of Africa like other self-respecting savanna lovers at 3.7 million years and 2.6 million years? In many respects it is silly to think that they didn't, or couldn't, when they had already populated most of Africa. Australopithecines were present by that time in the grasslands of Chad, Central Africa, 2,500 kilometers away from the Rift Valley of East Africa, where early hominid research has so far concentrated. Since similar savanna grasslands extended all the way from West Africa to north China, there were no impediments to much wider dispersals throughout this vast transcontinental grassland recently dubbed "Savannahstan."

By 1.8 million years ago hominids had reached Dmanisi in Georgia. The Caucasus Mountains, at the time, were not fully formed and the environment was still a warm Mediterranean type with many rivers and lakes, open woodlands, and forests on the riverbanks and slopes of the mountains with their deep valleys and ravines, rich in edible plants and animals. At Dmanisi hominid remains were associated with those of other animals that were also expanding their range, such as sabre toothed cats, other big cats, bears, horses, rhinoceroses, ostriches, rodents, hamsters, hyenas, antelope, gazelles, giraffes and gerbils.

So in the context of information on past environments, let's follow the imaginary dispersal of *Homo ergaster* out of Africa at, say, 1.75 million years ago. Moving unimpeded by great desert barriers, initially toward Israel, "Ergaster" would immediately be confronted by local hominids, since there are stone artifacts in Israel, at Erq-el-Ahmar, dated by paleomagnetism to between 2.0 and 1.8 million years. Moving farther north, "Ergaster" would have met the small Dmanisi hominids in their warm Pleistocene cul-de-sac.

"Ergaster" also could have walked woodland and savanna environments to reach the fertile Indus Valley and the richly wooded foothills of the Himalayas, where stone tools dated to at least two million years indicate that hominids had already moved in. Continuing farther east and south, and by now 8,000 kilometers from Africa on the east coast of Asia, "Ergaster" may have met another smaller hominid species, one that resembled the small Dmanisi people and later gave rise to an endemic little people on Flores. Finally, at the very southeast corner of the Asian continent, "Ergaster" would have reached Java to find an established population of large-bodied hominids with some very different, and more primitive, traits.

Homo ergaster clearly was not the only player on the Asian stage in the early Pleistocene. On their march east, would they have met other hominids traveling west? Was Asia always a passive recipient of hominids? Two-way traffic is well documented in other animals moving into and out of Africa. There is really no reason at all to assume that hominids always moved out of Africa into Asia. Given the lack of an obvious East African ancestry for *Homo ergaster*, could the species even have originated in Asia, evolved from one of the other hominid populations that had earlier migrated out of Africa? This would explain the sudden appearance of *Homo ergaster* in Africa better than invoking a short, undocumented process of evolution. In fact, the Dmanisi hominids may be "close to the stem" from which the *Homo ergaster* lineage evolved, and at a time when the core of grassland expansion was Asia—and Africa was peripheral. The expansion of savanna, with a corresponding need for people to cover greater distances and place more reliance on hunting, may have provided the impetus for their increases in stature and brain size that occurred about 1.8 million years ago. If so, was Asia the epicenter for some major developments in the human story? Maybe the little hominids from Dmanisi and Liang Bua, with their pre-*ergaster* traits, are the first real challenges to the "Only Out of Africa" orthodoxy.

The possibility was intriguing. We don't know which one of the early hominids was LB1's ancestor, but that's no surprise, given that we still don't know the full range of hominids that inhabited Africa and Asia. Even in East Africa several new groups of hominids have turned up in the past decade, as well as the first-ever fossil evidence of that well-known relative of ours, the chimpanzee—as if to emphasize that we're merely scratching the surface.

I felt this significance of LB1 would be lost if Peter insisted on naming LB1 *Sundanthropus,* and evidence that LB1 ought to be called *Homo* also kept surfacing. In January 2004, Chris Turney produced the first absolute dates for the LB1 skeleton based on associated charcoal samples. Previously, we had obtained an ESR date from Sector IV suggesting that the deepest deposits there might be about 100,000 years old, but given the characteristics of the LB1 skeleton, we anticipated that the associated deposits in Sector VII were going to be much older—probably beyond the age range of radiocarbon dating.

Chris's e-mail warned me that his dates were not what I was expecting. In fact, you could have knocked me down with a trowel. LB1, it appeared, had died just 18,000 years ago. Chris also confirmed the date for the white tuffaceous silt associated with the extinction of *Stegodon* and the little hominids. It was only 12,000 years old. There was nothing like this anywhere else on the planet! But I thought, what the hell, basically everything about the find was already wrong. LB1 only died 18,000 years ago, but in brain size and estimated height she most resembled the 3.2-million-year-old Lucy. How on earth could this tiny, small-brained hominid have made tools, hunted *Stegodon* and migrated to Flores across Wallace's Line, then survived in isolation for almost a million years, long after its African kin had disappeared? Even more extraordinary, how could it possibly have coexisted in the region with modern humans from at least 50,000 to 12,000 years ago?

I looked for error. Could the LB1 skeleton have been washed into younger deposits? Not possible, because parts of the skeleton were articulated. Besides, the stratigraphic layers exposed in the excavation were very clear and defined. No complex mixing process of old and younger materials had occurred. Anyway, the luminescence ages provided by Bert Roberts and Kira Westaway for sediments above and below the skeleton were entirely compatible with the radiocarbon results. After recovering from the shock, I realized that the dates just had to be right.

Peter at first refused, point-blank, to believe them. He said that if a spacecraft had landed in a rice paddy and deposited an alien, it would have been easier for him to accept. Actually, I thought the dates were one more good reason to consider LB1 as *Homo*. Peter, however, was as immovable as a block of Liang Bua limestone, and he wanted a check on the radiocarbon results. This was going to delay the papers being submitted to *Nature*. Normally a non-drinker, I had the very strong urge to have a Bintang beer, pretty much the only easily available brew in Indonesia. It would go very well with the packet of sweet-tasting clove cigarettes I was now smoking every day.

At least our first excavation season at Song Gupuh, a deeply stratified rock shelter in East Java, was going well. At this site the same team that worked at Liang Bua used shoring and similar excavation strategies to dig down 16 meters, without encountering bedrock—and obtained a cultural and faunal sequence spanning around 70,000 years. Song Gupuh has the same potential as Liang Bua and provides comparative information from an island on the continental side of the Wallace Line.

During the Song Gupuh excavation, we were based in two adjacent houses in the provincial town of Punung, where the traditional East Javanese food was excellent and the people very friendly. The houses were owned by an extended family, but I never got all the relationships figured. In one house, there was Pak Teguh, largely immobilized with a leg broken in a motorbike crash, his wife, Ibu Jumirah, their daughter with husband and two young children, and an aged grandmother; in the second house there were at least three other young women with children, a couple of young men, and several servants. Somehow archaeologists and the family managed to go about their business under the same roof with minimal disruption.

Teguh is particularly interested in birds and had at least eight renowned songsters in fancy cages distributed around the two-house complex. These birds were expensive—he had purchased the property on which the houses stood about 25 years previously with the money from the sale of just one bird, then financed the construction of his large house with another. Not bad for a man who started off as an orphaned kid from a poor background. His birds were good but not national singing champions, which can sell for $15,000 each—sometimes more—but he still has aspirations. Birds have been a major factor and passion in Teguh's life—just as archaeology has been in mine.

Inside, the houses could be pleasantly noisy, with people eating, sleeping and socializing, birds singing, and the clatter and splash of domestic chores. Just outside was the Punung main street with vendors of farming tools, clothing, music, fruit and vegetables, all vying for customer attention together with trucks, buses and vans bringing animals, produce and local farmers in their conical hats to and from the market. Punung is a traditional East Javanese town and foreigners are rare enough to be a real focus of curiosity. In fact, the only foreigners who visit Punung are those associated with archaeological research—Song Gupuh is only 15 minutes by car from the town.

At the same time, Aziz and his team from GRDC were in the field carrying out a survey of the river terraces along the Kendeng Hills section of the Solo River in East Java. For this they had contour-mapped the entire area using facilities in the GRDC Remote Sensing Lab to electronically overlay digital aerial photos and topographic maps. Their aim was to describe and date all the river terraces, and so reconstruct the history of river downcutting and landscape evolution. This would provide far more meaningful age estimates for vertebrate fossils in the area, including the remains of late *Homo erectus* previously excavated from the Ngandong river terrace.

Meanwhile, I went to a small village way up in the mountains of west-central Sulawesi to assist Harry Truman Simanjuntak from ARKENAS with excavations at Mananga Sipakko, an early Neolithic site on the banks of the Karama River. The Karama is a large river and runs at terrific speed after rain, which happened every day, it seemed. To get there the excavation team from ARKENAS, Balai Arkeologi Makassar and UNE, had to squat in a dugout outrigger equipped with three engines. It was like whitewater rafting without the helmets and safety vests. But well worth it, since the earliest pottery at this dig is very similar to the earliest Austronesian pots known from Taiwan, dated to about 5,000 years ago. There was also a lot of obsidian, and from previous research it was likely that some was traded from the Talasea Peninsula on New Britain 3,000 kilometers to the east in Melanesia, evidence of very long distance transport and cultural connections, and earlier than expected. Liang Bua and its little hominids were not the be-all and end-all of our "Astride the Wallace Line" project—or so I told myself!

But word about the discovery was now starting to filter out, and at the beginning of March, Leigh Dayton, a science reporter with *The Australian* newspaper, sent me an e-mail asking if the amazing rumor that we had found a 100,000-year-old australopithecine on

Flores was true. This was more disconcerting than it sounds because of *Nature*'s policy of having an embargo on any information relating to its forthcoming papers. If this was broken, our paper could be turned down.

At any time, Jacob, too, could turn up at ARKENAS and, with the support of Soejono, demand to see, study or even take LB1 back to his lab in Jogyakarta. He then only had to make the first public announcement in the *Jakarta Post* or publish a brief descriptive article in the University of Gadjah Mada *Bulletin*, and whatever he decided to call the specimen would become its scientific name. By the conventions governing taxonomic precedence, to get the honor of scientifically naming LB1, his published description did not have to be the best, or even particularly good—it just had to be first.

Finally, in early March, all the dates came in, and the different techniques cross-checked against each other beautifully: LB1 had definitely died about 18,000 years ago. So we submitted two papers describing the find and its archaeological context to *Nature,* with LB1 placed in a newly erected genus, *Sundanthropus floresianus. Nature*, which fast-tracked the referee process due to the potential importance of the finds, had responses for us within three weeks. All referees were supportive, excited even. One reviewer, perhaps with Piltdown in mind, began by saying that "at first I thought that this was an elaborate hoax." Another said that this "is not just another incremental hominid find. Rather, this discovery could one day rank as one of the most significant finds made in the past 50 years." Great! That was as important as *Zinjanthropus,* Lucy and *Homo habilis*, which shook the world because they showed the antiquity, and complexity, of humanity!

Encouraging feedback indeed, but the reports also provided

lots of constructive criticism from people who that really knew their craft: they requested additional reading, more information on the context of the find, additional analyses on the skeleton and computed tomography (CT) scan data, but the most consistent comment from the referees was that the species should be allocated to the genus *Homo,* not *Sundanthropus.*

Some of the requests also included the need for more anatomical detail and photographs. So in April 2004, Peter Brown and I returned to Jakarta, and with the rest of the Liang Bua team set about cleaning, reconstructing, and recording parts of the LB1 skeleton again. Peter also needed X-rays of the skull and lower jaw in order to be able to examine the roots of the teeth and other internal features. With the benefit of an official letter from Dr. Tony Djubiantono, the new director of ARKENAS, Thomas Sutikna arranged access to the facilities at Fatmawati Public Hospital, but asked for computed tomography (CT) scans instead of the X-rays requested by Peter. The mix-up turned out to be very fortuitous. CT scans, sometimes called CAT scans, are normally used for diagnosing cancers, cardiovascular diseases, trauma and musculoskeletal problems of all sorts, and many shock or trauma centers and hospitals have a CT scanner in the emergency room. But what I did not realize at the time was that with new laser technology, CT data can also be used to make accurate 3-D resin replicas of scanned materials.

At Fatmawati Hospital, a curious crowd gathered in the CT scanning room while the remains of a 30-year-old woman who died 18,000 years ago were secured to a bench, which was then moved forward into the large, donut-shaped scanner. Most of us then retreated to watch from the safety of the adjacent control room; only Peter, wearing a radiation-proof vest, remained in the room to accompany LB1 while the scans were undertaken.

In the control room, two highly skilled technicians manipulated the CT scan data, and on a computer screen were able to create virtual copies of the skull and lower jaw, which could be rotated in any direction, be digitally sectioned at any point or in any plane, and be used to produce large prints on negative film. Peter was able to look at the multiple roots of LB1's premolar teeth and get his photographs.

However, one of the sticking points with revising the manuscripts was Peter's insistence that the genus name *Sundanthropus* be retained—that LB1 could not possibly be *Homo*—and that the species could not be the result of dwarfing because hominids with the insulating effects of culture were not subject to the same island evolutionary processes as other large mammals. In this he was at odds with my Indonesian colleagues, Bert Roberts and me. We were also going to strike problems with the referees, most of whom did not agree with Peter, and this would cause further delay. Despite often heated exchanges with Bert and me, Peter stuck to his guns. Then one morning he telephoned me in my office and said, "It's *Homo*." No other explanation was offered, but I didn't inquire further. We had edged that much closer to getting the manuscript back to *Nature* and getting into print.

With the problem about the genus name solved, there was still debate about the species name. In the first submission we had used "*floresianus*," but one of the referees had said that we risked generations of paleoanthropology students referring to LB1 as *Homo* "flowery anus," an indignity we did not want to bestow. Alternatives included *floresi, floresiensis, florescus, manggaraii* (after the local ethnic group in West Flores) and even *hobbitus*—referring to the extremely small size of the species, after an imaginary race of half-sized, hairy-footed characters in the universally popular *The Lord of the Rings* by J.R.R. Tolkien.

The name "Hobbit" was said to mean "hole builder" or "hole dweller," which also seemed appropriate given that LB1 had been recovered from a very deep hole. The main appeal of using it as species name, however, was the thought of learned types at conferences having to seriously discuss the attributes and evolutionary history of *Homo hobbitus*. In the end common sense prevailed, and in the second round of submissions to *Nature*, LB1 became the holotype specimen for a new human species, *Homo floresiensis*. Only later did we find out that *Nature* editor Henry Gee is a Tolkien buff. He has even written a book called *The Science of Middle Earth*. Henry might have liked the taxonomic reference to hobbits.

Peter, in the meantime, had sent me an unequivocal e-mail:

> As the referees note, the lunatic fringe are going to have a field day with the Flores midget. The last thing I am willing to do is pour fuel on that particular fire. This is what any reference to hobbits or midgets would do. I can't believe that you could be serious about this, Not funny, not a joke, it is hard enough to push this particular barrow with only one individual and nothing comparable from anywhere else. Certainly would not help you and Bert get a job. Everyone would think you had gone nutty.

He warned that any mention of hobbits in a press conference, even as a nickname, and he would walk out. This was a bit of a shame, really, because it was, I thought, singularly appropriate; a little person who that lived in a cozy hole in the ground on an isolated Middle Earth island. Hobbits were also familiar with a type of extinct elephant, and were chased by Komodo dragons, the Flores version of the fire-breathing dragon, Smaug. It was also not lost on me that the reason that the *Australopithecus afarensis* skeleton known as "Lucy" had such public recognition and staying power was because of her nickname, which was taken from the Beatles' song "Lucy in the Sky with Diamonds," fortuitously played as Johanson's team of anthropologists, high as kites with the euphoria of the discovery, celebrated into the night.

As if to emphasize the point, even older hominid finds, such as the spectacular six-million-year-old *Ardipithecus ramidus*, found in 1994, and *Sahelanthropus tchadensis*, found in 2002, at nearly seven million years old (double Lucy's age), after a big initial splash, faded from public view as quickly as they came. They did not have catchy nicknames.

As it transpired, the matter was out of Peter's hands; my younger Indonesian colleagues liked the name "Hobbit" and had begun to use it affectionately for LB1. "Hobbit" stuck. Now the nickname has even made it as a supplementary entry into the *Collins Australian Dictionary*—"a very small type of primitive human, *Homo floresiensis*, following the discovery of remains of eight such people on the island of Flores, Indonesia, in 2004."

Islands in the Evolutionary Stream

I took another look at the e-mail from Peter Brown. "Everyone would think you had gone nutty," he warned, and then continued, "Have you ever heard of Grover S. Krantz?" Peter had a point. The late Grover Krantz was an anthropologist and cryptozoologist (the study of hidden animals such as the Loch Ness monster, leprechauns and other little people), well known for being a controversial and outspoken authority on the North American Bigfoot, or Sasquatch. He had written several papers and four books on the topic, and pushed the theory that Bigfoot was a remnant population of the giant ape *Gigantopithecus* (which is in the same lineage as the orangutan, but which had gone extinct in the late Pleistocene), and was a possible candidate for tribal memories of the Yeti.

Many if not all cultures have stories of giants and little people. When visiting Flores just before our *Nature* publication, Bert Roberts and Gert van den Bergh went in search of local Nagakeo stories of little people, as recounted in Gregory Forth's 1998 book *Beneath the Volcano*. These amazing stories, common in central Flores, told of dwarf-sized folk, about a meter tall, with long hair, potbellies, ears that stick out a bit, a slightly awkward gait, and longish arms and fingers, and in females, very pendulous breasts that they apparently threw casually over their shoulders. Apart from the breasts—hard to vouch for with skeletal evidence—the descriptions fitted Hobbit to a T. The villagers said that the little people would murmur to each other and repeat villagers' words parrotlike. The villagers called the small folk Ebu Gogo, meaning "the grandmother who eats anything raw," including vegetables, fruits and meat—even humans. If food were served to them on plates made of pumpkin rind, they would eat the plates, too.

The raiding of crops by the Ebu Gogo was tolerated, but when one stole and apparently ate a human baby, it was too much, and a plan was hatched to rid the region of Ebu Gogo. The villagers collected dry palm fibers and followed the Ebu Gogo to their hideout, a cave in a cliff face on the slopes of Ambolobo. Using sticks long enough to reach up to the cave entrance, the villagers offered dry palm fibers to the Ebu Gogo, who greedily took in the offerings. But then the villagers threw a burning bale of palm fiber into the cave, turning it into an inferno. Singed, the surviving Ebu Gogo fled. They were last seen heading north just before the village moved location and not long before the Dutch settled in that part of Flores in the 19th century.

Were these creatures hobbits? Bert and Gert were at least partially convinced. Nutty? Maybe. Apart from not believing local folklore, most archaeologists will never have contemplated an island dwarfed version of *Homo*, such as *Homo floresiensis* or little people like Ebu Gogo, because hominids were meant to be largely insulated by their cleverness from island evolutionary processes that can dwarf large animals, or make giants out of small ones.

But at least one of the referees of our *Nature* papers thought that the idea of an island dwarfed human that evolved in isolation on Flores was not as fantastic as it sounded. This scenario was more likely, said the referee, than one in which *Homo floresiensis* derived from the early dispersal out of Africa of an australopithecine. The latter was just a little too fantastic. Nevertheless, likely or not, if true it would be the only time in all of human evolution that a human population was dwarfed to this degree. Had the extreme circumstances on the oceanic island of Flores really pushed the boundaries of the *Homo* genus? The evolutionary histories of other animals on islands indicate that this is precisely what happened.

Once animals are established on islands, they can evolve body sizes very different from those of their ancestral migrants. The "island rule," an idea first proposed by Bristol Foster of the University of British Columbia, states that mammals bigger than a rabbit, including mammoths, elephants, *Stegodon* and hoofed animals, may shrink to diminutive pygmies, while smaller mammals, as well as reptiles and birds, may get bigger. The shrinking of larger mammals occurs on islands with limited resources. In the absence of predators, smaller individuals will be advantaged by their reduced food requirements and shorter pregnancies. It is not surprising that the same selective pressures operated in much earlier geological times on other groups of large animals. For instance, sauropod dinosaurs, with a total body length of up to 45 meters, were the largest animals ever to live on land. But about 150 million years ago, a diminutive sauropod species only six meters long evolved on a large island in what is now northern Germany. Growth lines in the long bones of this dwarfed species indicate that downsizing was achieved by a decrease in growth rate from that of a larger ancestor.

Such are the evolutionary forces at work that insular dwarfing can happen very quickly, as illustrated by the rapid dwarfing of red deer on Jersey in the Channel Islands, 25 kilometers from the coast of France. These deer became reduced to one-sixth of their body weight in less than 6,000 years, during the last interglacial period around 120,000 years ago.

Often because of the absence of large mammalian predators and the paucity of resources, reduction in body size is accompanied by a reduction in the size of the brain, an energy-expensive organ to maintain, while bone fusion and shortening in the limbs result in more heavily built legs with stouter bones: speed is sacrificed for increased stability. In addition, animals on islands commonly develop more efficient ways of chewing and digesting food.

The energy usually used to outgrow, outrun, outfly or outthink predators is instead channeled into specific, occasionally extraordinary, island adaptations. Birds on oceanic islands, for instance, time and again lose the power of flight. There is no need for such an energy-sapping adaptation in an environment free from ground-dwelling mammals. Instead, birds squeeze into niches occupied by large browsing mammals on the mainland by supersizing, and shifting diets. New Zealand moas, which included the tallest birds ever, are a prime example.

Elephants were the most characteristic and common large mammals on Pleistocene islands in both the Old and New Worlds, where they developed many of these insular traits; for example, on Sardinia, Sicily, Malta, Crete, Cyprus, and the Greek Cyclades and Dodecanese islands in the Mediterranean. Here they quickly dwarfed to pygmy size—one as small as 90 centimeters—to teeter about on the scrubby, sun-washed Mediterranean slopes, the modern-day abode of goats and sheep. On the mainland, elephants dominated other competitors, or avoided predators, by outgrowing them. But on many resource-limited islands, where the main

Endemic fauna on Malta included pygmy elephant, pygmy hippo, giant tortoise and a large goose. These animals well illustrate the unbalanced nature of island faunas and the size changes that often occur. (FROM ANDREW LEITH ADAMS, *NOTES OF A NATURALIST IN THE NILE VALLEY AND MALTA*, 1870)

competitors were other individuals of their own species, different elephants followed the same, parallel evolutionary patterns; predictable patterns that included dwarfing, changes in the skull and teeth, and shortening of limbs.

Perhaps the most famous case of island dwarfing is the cave goat from Majorca, *Myotragus*, a survivor of the time when the Mediterranean was completely drained of water, six million years ago, due to the collision of Africa with Eurasia, ramming the

Straits of Gibraltar shut. Finally, about five million years ago, the tectonic forces relaxed and, combined with a global sea level rise, the ocean poured into the Mediterranean Basin. Miraculously, *Myotragus*, then a mountain goat, survived, as its mountain stronghold became an island surrounded by surging seas.

Specializing on tough, limited foods, *Myotragus* dwarfed in body mass by at least 60 percent, and its teeth changed for increased feeding efficiency—the molars developed higher crowns and cusps, the canines were lost, and the incisors lengthened and became ever-growing like those of rodents. Remarkably, no other animal landed on Majorca to compete with it for food, and no carnivore turned up to eat it. Without the need for vigilance, flight or the early detection of predators, the eyes of *Myotragus* migrated from the sides of its head to the front for stereoscopic sight, like ours, and there was a 50 percent reduction in brain size that most affected the areas associated with vision. It also abandoned the need for rapid running and instead developed short, stout legs for slow-gait, small-step stability. All these traits made for more efficient energy use, and therefore increased the animal's fitness for living in predator-free, resource-challenged conditions. But five million years of evolution disappeared as soon as modern people arrived on the island about 10,000 years ago, and *Myotragus* quickly became extinct.

Although there are many general trends in island evolution, some morphological traits evolved in response to very specific circumstances. Gargano, now a peninsula on the Italian coast, was an island in the late Miocene about five million years ago, and had a population of an antelope-like animal, *Hoplitomeryx,* which developed five horns, including one in the center of the head and one above each eye. This peculiar arrangement may have arisen to combat their only predators on the island, which lacked protective

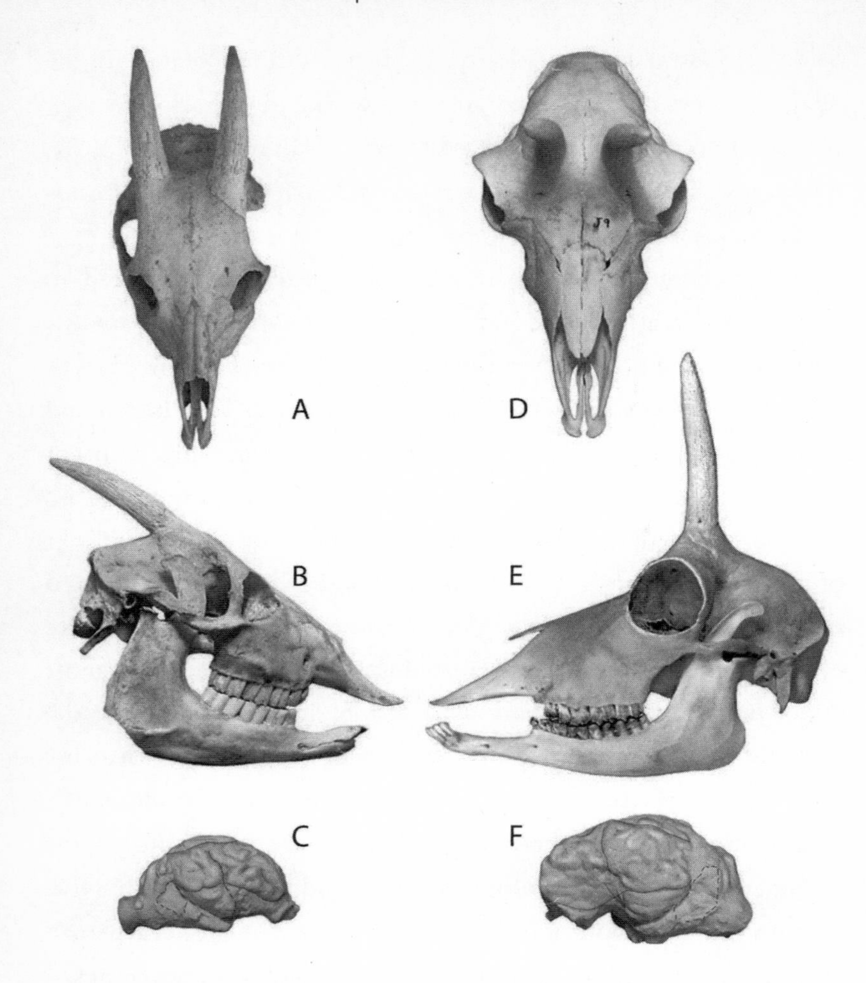

The skull (A, B) and brain (C) of Myotragus, the endemic cave goat from the island of Majorca (left), compared with an equivalent species from the European mainland. During five million years of isolation, Myotragus' body and brain shrunk and its eyes migrated from the sides of the head to the front. (CREDIT: MIKE MORWOOD, AFTER KOHER AND MOYÀ-SOLÀ, 2003)

cover. Young animals were open to attack by giant buzzards during the day and by giant owls at night, and both birds probably first tried to strike the eyes or back of the neck of intended prey.

For all its seemingly weird outcomes, in fact, much of island

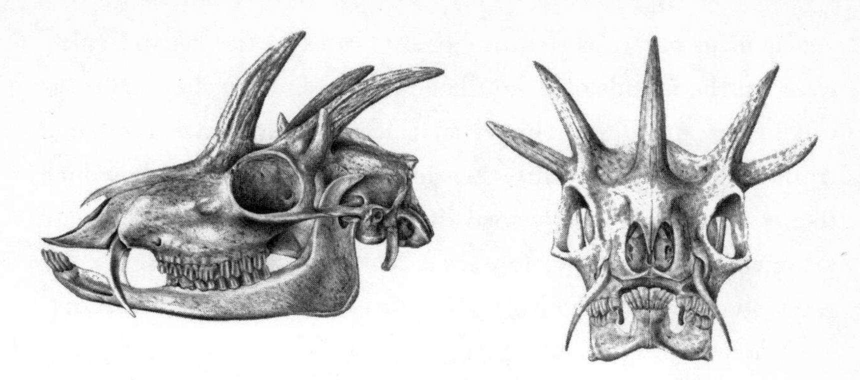

Hoplitomeryx, *the antelope-like animal from a (former) island off the coast of Italy, developed five horns to combat the only predators on the island, giant owls and buzzards.* (FROM LEINDERS, 1984)

evolution can be explained as predictable responses to a resource-limited, albeit radical environment. Small animals, ecologically released from the selective pressure of large competitors and predators, become big. Large animals, in contrast, released from the pressure to outgrow competitors, dwarf to a size best suited to meet the challenges of survival on a specific island. Not surprisingly, however, the size of an island relative to the size of the evolving species greatly influences the course evolution will take. If an island is big enough with sufficient resources and habitat diversity, one animal species can split into several that exploit the different habitats, a process known as adaptive radiation. Even small islands, such as the Galápagos, can produce a profusion of species if the evolving animal is small like a finch.

Not every small mammal, reptile or bird that reaches an island will increase in size, but in the course of adaptively radiating to fill empty habitat niches, some do. This is clear in the case of the West Indies, where rodents radiated spectacularly. The resulting species included A*mblyrhiza,* the world's largest rodent, the size of an American black bear with a weight of up to 210 kilograms.

Amblyrhiza, one of the most extreme cases of the "island rule," lived on the islands of Anguilla and St. Martin in the northeastern Lesser Antilles. A hypervariable species, its size oscillated from 60 to 210 kilograms, tracking the size of its islands, which themselves waxed and waned in size from 150 to 2,500 square kilometers throughout more than 20 Pleistocene glaciations. Always slightly out of phase with the islands' oscillations, *Amblyrhiza* was caught out by the very high sea levels of the last interglacial period. A giant of glacial times, it was too big for the abrupt (perhaps less than 500 years) and catastrophic habitat loss that accompanied the last interglacial, which shrunk its home to less than 10 percent of its former extent. Unable to change body mass quickly enough, *Amblyrhiza* became extinct a little before 100,000 years ago.

Larger animals require larger islands with more abundant and diverse resources to be freed from the constraints of the island rule and for adaptive radiation to occur. This is seen in the case of Madagascar, a very old, truly large island that became an evolutionary center in its own right.

Madagascar has been isolated ever since it broke away from the southern supercontinent of Gondwana and then drifted south to reach its present position 430 kilometers off the East African coast about 130 million years ago, well before mammals evolved. Gondwana animals that managed to survive and evolve on the island until recent times include elephant birds—huge flightless relatives of the emu, ostrich, rhea and moa. The island has always been difficult to colonize. Mammals that managed the feat then evolved *in situ*, include hippos, civets, mongooses, catlike carnivores and lemurs, with most being early offshoots from their evolutionary

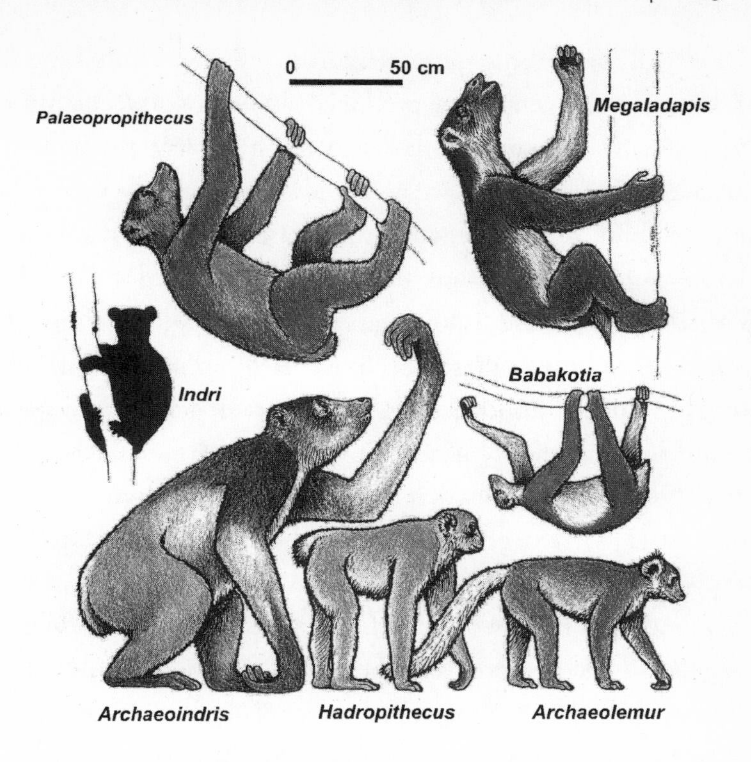

The many species of lemurs on Madagascar arose from a single colonizing species that made the crossing from Africa about 60 million years ago. Elsewhere lemurs, an early and formerly widespread relative of monkey and apes, subsequently became extinct. (CREDIT: MIKE MORWOOD, AFTER MITTERMEIER *ET AL.*, 1995)

lines. Lemurs in particular have undergone spectacular adaptation and radiation, with the many species each occupying a specific habitat from the island's wide array, and of great variability in size and shape—including a now extinct, giant form about the size of a gorilla, and others that looked like sloths, koalas and monkeys. However, DNA evidence shows that all evolved from a single colonizing species that made the transmarine crossing from Africa. Other major groups on the island, such as the insectivores, rodents and carnivores, also seem to have each radiated from single species arriving from Africa.

It is really something quite extraordinary: a handful of successful colonization events have produced a whole world, the miracle that is the Madagascan fauna. As a result, nearly all the native mammals of Madagascar are endemic, meaning they evolved on the island and do not occur naturally elsewhere. Little disrupted this Madagascan profusion until humans arrived from island Southeast Asia around 2,300 years ago. Hunting, forest fires and clearance then led to a drastic decline in large animals within a few centuries. But the subsequent disappearance of at least two species of giant tortoise, three species of pygmy hippo, two species of elephant bird and 17 species of lemur occurred as a drawn-out "stepwise cascade of extinctions" rather than as a single blitzkrieg event—and the process is continuing.

Animals that have evolved in largely predator-free, island isolation generally do not fare well when people arrive, and this vulnerability to humans turns out to be another characteristic of all island faunas. In fact, giant tortoises, which were common in island faunas worldwide, may be more symptomatic of human-free rather than island environments specifically. Although giant tortoises are often quoted as an example of reptiles increasing in size on islands, they are more likely to have been large on arrival. All the large species of tortoise formerly present on Indonesian islands, for instance, were very similar to a giant tortoise found in the Siwalik fossil deposits of continental India that accumulated during the Miocene—a geological epoch between 23 and 5.5 million years ago, when the world was wetter and warmer. There is heat-retention advantage in cold-blooded tortoises being large, while their size and ability to float for extended periods without fresh water or food gave them a major advantage in island colonization—hence their presence on the Galápagos Islands some 1,000 kilometers from the nearest source area, the South American mainland. With their shell armor, giant tortoises are effectively immune to predation—except by humans—and in all documented cases, such as on Cuba and Madagascar, their extinction has coincided with our arrival. Remote areas, such as the Galápagos Islands, where giant tortoises still survive, just happened to be their last human-free refuges. Their disappearance, such as occurred on Java about 1.2 million years ago, may therefore be evidence for the presence of tool-making hominids.

Spectacular evolutionary changes and the survival of early evolutionary lines, such as lemurs in Madagascar and marsupials in Australia, occur on islands that were largely cut off from further colonization. One of the main causes of such isolation appears to be ocean currents. For instance, the islands of the Balearic Archipelago, including Majorca and the Pityusic Islands, had very different faunas, stranded when the sea reinvaded the Mediterranean Basin about five million years ago. The Pityusics were occupied by a fauna comprising a giant dormouse, giant tortoise and 20 species of snail. All of these went extinct about two million years ago and were never replaced! That no land mammals, reptiles or land snails were able to recolonize these islands is due partly to their distance from the mainland, about 90 kilometers off the Iberian Peninsula, but more so to the pattern of marine currents in this part of the western Mediterranean that sweep past the Balearic Archipelago with enough strength to keep rafting animals from its beaches. The current-induced isolation was only breached with the arrival of modern humans—not from the adjacent Iberian Peninsula, from where the Balearic Islands are sometimes visible on the horizon, but by seafarers sailing from the Rhone River delta area about 450 kilometers to the north in what is now France.

Whether they swim or raft on flotsam, land animals are at the mercy of ocean currents when making sea crossings. Currents will either assist or prevent their reaching an island. Until humans invented sails and keels for their watercraft they were similarly constrained.

With the exception of animals such as *Myotragus* on Majorca and elephant birds on Madagascar, which have taken advantage of land bridges formed at times of low sea level or which occupied mainland areas that later became islands, colonizing animals have to make water crossings, either deliberately or by accidental "sweepstake" dispersal. Just a tiny minority of animals are good at this; mammals show a remarkable incapacity for colonizing islands.

The supertramps among mammals are usually bats and rodents: the latter constitute nearly half the species of mammals currently alive today in the world, can ride and survive on flotsam, and if cast up on an island can reproduce rapidly. Large-bodied animals that reach islands will usually be good swimmers that on the mainland often inhabit swampy places, or places at least seasonally inundated. In the case of mammals, these include elephants, hippopotamuses, pigs and deer. The fact that all these animals tend to travel in social groups and are herbivores, or omnivores in the case of pigs, also gives them an edge in successfully colonizing islands. They are much more likely to make crossings in biologically viable groups and find enough food to survive. But even among these species the number of successful colonizers, those that actually make it to an oceanic island and survive long enough to mate and create a sustainable population, is vanishingly small. How small? Only in the Philippines, adjacent to Indonesia, has this dynamic process been investigated in any depth.

An archipelago of over 7,000 islands of widely differing sizes and history in close proximity, the Philippines comprises some islands that are oceanic; some that are fragments of once larger islands; and some, such as Palawan, which periodically had connections to the Asian mainland via Borneo. Not surprisingly, the oceanic islands in the Philippines with the greatest number of species and degree of evolutionary divergence are also geologically the oldest. For instance, Luzon, which is about 32 million years old, has 26 terrestrial mammals, of which 20 are endemic. In contrast, the Sulu islands, formed within the last two million years, have only seven native mammal species of which two, a shrew and a rodent, are endemic.

Given the number of endemic and nonendemic species, and the ages of the various oceanic islands, it is possible to calculate the rate at which successful over-water colonization occurs: about once every half-million years, and that for supertramps crossing distances of no more than 15 kilometers. What are the chances for other animals making it: one in a million? One in two million? But given long enough, some animals can still beat incredible odds. It is still difficult, for instance, to explain how Old World monkeys managed to cross the Atlantic Ocean from Africa to reach South America about 35 million years ago, and so give rise to New World monkeys. But it happened.

Past and present distributions of animals on islands provide a measure of their dispersal ability, the difficulty of specific crossings and the sequence of colonization. In the case of large-bodied land animals in Southeast Asia, giant tortoises and *Stegodon* are the champion colonizers, being found in fossil records on islands throughout the region. Most other large-bodied animals were more restricted in their distributions. This is clear when the fossil records from a north to south transect of islands in the region, from the Philippines down to Timor, are compared.

The earliest land animals known from the Philippines, on the island of Luzon, comprise giant tortoise, a pygmy *Stegodon*, a pygmy elephant, a peculiar species of pig with four tusks and distantly related to African warthogs and rhino. Around the same time, 2.5 million years ago, the same suite of terrestrial animals occurred in Sulawesi—giant tortoise, a pygmy *Stegodon*, a type of pygmy "elephant" with four tusks and African-type molar cusps, and a peculiar species of pig with four tusks. Only rhino failed to make the crossing.

Over time, isolated and undisturbed, these animals show evidence for further adaptation to their island conditions: the number of ridges on the molars of pygmy *Stegodon* increased (i.e., became

(Left) *Animals on Sulawesi illustrate the unbalanced nature of island faunas and the size changes that often occur.*
(Right) *Celebochoerus heekereni, a primitive pig found in Sulawesi around 2.5 million years ago.* (CREDIT: HANS BRINKERINK)

more hyperdont), probably as an adaptation to tougher foods such as grasses, while the pigs developed shorter legs and became smaller: food was restricted and in the absence of predators there was no need for speed. By 840,000 years ago, however, all these island-adapted animals had disappeared, to be replaced by large species of *Stegodon* and elephant. Intriguingly, the presence of stone tools in the same strata indicates that hominids may have arrived on Sulawesi and Luzon around the same time, and on both islands this resulted in the extinction of giant tortoise, pygmy *Stegodon* and pygmy "elephant."

Farther south, Flores had far fewer large land animals—only giant tortoise and *Stegodon* made the passage—but a remarkably similar change in animals occurred there by 840,000 years ago, when pygmy *Stegodon* and giant tortoise became extinct, and large *Stegodon* recolonized the island. In this case, the turnover in animal species was definitely associated with stratified, well-dated stone artifacts that constitute indisputable evidence for the presence of hominids. Farther still to the southeast, there is undated evidence that the same replacement occurred on Timor, where the indomitable Father Theodor Verhoeven found evidence for pygmy *Stegodon* and giant tortoise in deposits predating those containing large *Stegodon* and stone artifacts.

The fossil evidence suggests that there was a dispersal of animals south through the oceanic islands of Southeast Asia and a corresponding drop-off in the range of animals that succeeded; that the Philippines was the main source for animals reaching Sulawesi; and that sea crossings from Sulawesi to Flores and subsequently Timor were difficult and infrequent. This all makes sense in terms of island geography and the predominantly southerly direction of ocean currents in the region. While the main cluster of Filipino islands was never connected to the Asian mainland, the required water crossings to reach them would have been narrowed considerably at times of low sea level. Ocean currents and stepping-stone islands would then have assisted the dispersal of animals from the Philippines south to Sulawesi.

Claims that stone artifacts occur together with the remains of large-bodied *Stegodon* and other archaic species in the Philippines, Sulawesi and Timor obviously need to be checked. But if the claims are substantiated, then early hominids will have to be added to the short list of large-bodied animals that are champion island colonizers. Indeed, the archaeological evidence from Flores indicates that we should add early hominids to the list now.

How would early hominids have colonized the islands east of the Wallace Line? Perhaps they floated across on rafts or clinging to vegetation swept down the major rivers gushing out of a larger ice-age catchment into the sea off mainland Asia. We know this can happen. In the devastating tsunami of December 2004, a woman who was seven months pregnant drifted 100 kilometers in seven days on floating vegetations. Two months after being saved she gave birth to a healthy baby. Over hundreds of thousands of years, many hominid individuals, and occasionally groups, must have been swept out to sea, and some of these hominids must have been swept ashore. If hominid colonization of island Southeast Asia occurred accidentally, then dates for the initial arrival of hominids on these islands should decrease significantly from north to south. My guess would be that hominids arrived on Sulawesi long before a small group was somehow washed out to sea, carried south by currents and deposited on the north coast of Flores.

Relatively few animals reach islands, but once ashore, established and given unmolested opportunity long enough, they can evolve into new species and adaptively radiate to a spectacular degree, as occurred on Madagascar. In island Southeast Asia, Nusa Tenggara and the Moluccas together make up a major center of global bird diversity, having more endemic birds than anywhere else on Earth, albeit dominated by a relatively small number of families. Alfred Russel Wallace dealt with such dizzying diversity in the same way that Charles Darwin dealt with the diversity of finches and giant tortoises that he observed on the Galápagos Islands—by inventing the theory of evolution to explain the evolution of new species and adaptive radiation. In Nusa Tenggara and the Moluccas, Wallace clearly saw that the ancestors of the birds and mammals came from geographically nearby lands. Which is what you might expect, if the geographical distribution of animals echoes their evolutionary history. Sulawesi, however, stumped him because it is a lost world of strange, "relict" animals, such as *Babirusa* and anoa—species with no obvious relatives; species that indicate how little we actually know about the early history of animal dispersal and evolution in Asia.

Babirusa, literally "pig-deer" in the Indonesian language, is an extraordinary barrel-bodied creature resembling a pig but with deerlike long and slender legs, and curved canines resembling horns that grow up through the snout and curve back toward the eyes. As Wallace noted, "the *Babirusa* stands completely isolated, having no resemblance to the pigs of any other part of the world." *Babirusa* has a complicated stomach, unlike other pigs, which have simple stomachs, and it also has a relatively unspecialized snout that it does not use to root: instead the animal feeds on fallen fruit and grubs. Its muscular structure, too, bears more of a resemblance to African warthogs, while its naked appearance gives it the cast of a hippopotamus. In fact, its nearest relative seems to be a 12-million-year-old ancestor of the hippo, *Anthracotherium*, found in the Siwalik fossil beds of northern India—but also found (rarely)

The skull of a Babirusa, *an early offshoot of the pig family that is only found in Sulawesi and adjacent islands. The "horns" of the animal are actually the canine teeth that grow up through the snout.* (FROM WALLACE, 1883)

in two-million-year-old deposits on Java, and layers of unknown age on Timor.

As for the sapiutan (literally "forest cow"), or anoa, Wallace couldn't work out whether it was an ox, buffalo or antelope. There are, in fact, two species—the Lowland Anoa, which is around one meter at the shoulder, and the Mountain Anoa, which is about 75 centimeters, and in many respects they seem to approach some of the oxlike antelopes of Africa, as well as the Pliocene aged *Hemibos* from the Siwaliks. Recent genetic research suggests that this dwarfed buffalo is one of the most primitive of living cattle, with ancestral characteristics harking back to the antelope roots of wild cattle. Possibly for more than five million years, since the late Miocene epoch, anoa has been able to survive and speciate relatively unmolested on the island ark of Sulawesi.

What about tarsiers, some of the world's smallest primates?

Anoa, a primitive bovid found in Sulawesi. The two known species are the smallest-known bovids and hark back to the oxlike antelopes of Africa. They have no extant relatives on the Asian mainland. (PHOTO: COLIN GROVES)

They now occur in Borneo, Sumatra, the Philippines and Sulawesi, but are the only living descendants of an ancient group that was once widespread in Eurasia and North America. Tucked away in the refugia of island Southeast Asia, they have survived their relatives by more than 30 million years with only minor changes in dental pattern and shape.

None of Sulawesi's six species of squirrel have nearby relatives either. Genetic evidence points to a single founder event for all Sulawesi's squirrels, resulting in explosive diversification in the island's empty niches. Using a molecular clock, it appears that the Sulawesi colonization event coincides with the lowest pre-Pleistocene sea level known, at around 11 million years ago.

The great majority of mammalian species on oceanic islands, however, are rats and mice. And Sulawesi is no exception. Here, a rapid evolution of rodents began around six million years ago. Today there are around 40 known species on the island of which some have no close relatives on the Asian mainland, their either having gone extinct or evolved into different forms. Others are related to some of the most primitive rats on the continental islands of Java, Sumatra and Borneo.

Overall, Sulawesi has swept up representatives of seven of the 29 placental mammal families (not including bats and rats) found on the Sunda Shelf. What about from the other direction—from Australia? There are two native Sulawesian marsupials that also happen to be the most primitive of the entire cuscus family. Restricted to Sulawesi, and a few nearby islands, the bear cuscus is a striking and distinctive animal with short, coarse black fur variably tipped with yellow. It has a bearlike snout, and is the only cuscus with round pupils—suggestive of daytime activity. With large feet and long limbs, it is a very un-possum-like possum. Indeed, because so many of the features of this animal are so primitive, it has been put into its own subfamily. The small Sulawesi cuscus, a small plain-colored animal that looks a little like the Australian brushtail possum, is also restricted to Sulawesi with a close relative on two offshore islands.

Given that marsupials have an almost total incapacity to colonize islands of their own accord, the success of Sulawesi's marsupials must rate as one of the longest odds in the evolutionary history of the region. In fact, the latest genetic evidence points to a fascinating scenario for the Sulawesian cuscuses that involved rafting on microcontinental fragments that sloughed off the margins of Australia as it moved northward. By 25 to 20 million years ago, these crustal fragments began to drift westward toward Southeast Asia. Some eventually collided with western Sulawesi, which lay at the very margin of the Asian continental plate. Clinging to these were, it seems, the ancestors of Sulawesi's cuscuses, newly diverged from the Australian possums and now stranded on a newly emerged proto-Sulawesi.

Babirusa, *Sus celebensis*, anoas, the bear cuscus, the small Sulawesi cuscus, all six species of Sulawesian squirrel, at least some Sulawesian shrews and most Sulawesian rodents—all are primitive families in their own right, or primitive members of their families, insulated from the tides of change that subsequently swept across the Asian mainland. Sulawesi is an old, large island—about 190,000 square kilometers. Moored 120 kilometers off one of the most biologically complex regions in the world, it has received and nurtured the animal flotsam of eons, allowing different groups from different epochs to survive side by side, and often for millions of years after their continental counterparts had been replaced.

And what about Flores immediately to the south?—not nearly as old as parts of Sulawesi, or as large, and always a more difficult colonization prospect. It was beyond the dispersal capabilities of elephants, pigs, deer, bovids, squirrels, civets, shrews, tarsiers and monkeys. But Flores is old enough—the evolution of some endemic rodent species began around eight million years ago—for two types of large-bodied mammals, *Stegodon* and hominids, to have reached the island by 840,000 years ago. Flores was also large enough with sufficient resource levels and diversity in several extensive environmental zones for both these early arrivals to become established, survive long term and evolve into new species.

The evolutionary history of animals on Flores then followed many predictable trends. In isolation *Stegodon* on the island dwarfed and developed additional ridges of their molar teeth for more efficient chewing of coarse foods, while some rodents increased in size. The small number of animal species on the island also allowed niche-filling opportunities. Most notably, apart from pythons on Sulawesi that are up to 10 meters in length, Komodo dragons found on Flores and the nearby islands of Komodo and Rinca are the only large, terrestrial, vertebrate carnivores in the world inhabiting relatively small islands. They can grow to more than three meters long, and as hunters as well as scavengers can kill animals as large as adult water buffalo and deer, can sprint fast enough to catch monkeys on the ground and can fish. It is likely that ancestral Komodo dragons, as with the giant tortoises, were already large when they arrived on Flores. Similar animals are found in the fossil deposits of the Siwaliks, as well as on Java and Timor—and a larger version, *Megalenia*, that was up to six meters in length, occurred in Pleistocene Australia. Komodo dragons were able to fill the vacant niche as top predator, and survive long-term on islands that would not support large mammalian carnivores, because they have about 1 percent of the calorific requirements of equivalent-sized mammals.

Although Hobbit is a strange hominid, she and her kind have many of the attributes that might be expected on the basis of what has happened to other large mammals on relatively small islands. She is small; has shortened, stoutly built lower limbs, sacrificing speed for increased stability; and was probably adept at tree climbing, as indicated by the unique structure of the shoulder and upper arm, long and thickened arms, long feet, and curved toe and finger bones. Whether the long and thickened arms, short legs, flared pelvis and overall australopithecine-like body shape of *Homo floresiensis* result entirely from secondary evolutionary convergence or reflect some of the traits possessed by the ancestral population is not yet clear. But the extraordinary length of the feet must represent an adaptive response to very specific environmental circumstances on Flores over a very long time, as there are no parallels in any other known hominid species.

Some of the means by which *Homo floresiensis* got around were obviously very different from those of other members of genus *Homo*, while jaws and teeth of the species are designed for superchewing, presumably as an adjustment to the tougher foods available on an oceanic island. On Flores, with its limited resources and few competitors, endemic hominids seem to have physically, and therefore behaviorally, adapted to new habitat opportunities.

Hobbit's brain also underwent the most extraordinary evolution yet seen in any hominid. While it retained primitive traits, it had uniquely enlarged frontal and temporal lobes—precisely those areas concerned with higher cognition activities, such as initiative and planning. Her brain may have been small, but it was normal, and the front part was restructured in a way that meant LB1 was probably smart, had language and could plan ahead.

The actual magnitude of the changes evident in LB1 was unanticipated, but those changes have to be seen in the context of at least 830,000 years of hominid isolation and adaptation—well over 40,000 generations. The same process of insular evolution, but over a much shorter time scale, is evident in modern human remains such as those that Father Verhoeven excavated from a number of rock-shelters on Flores. The 3,550-year-old skeleton of an adult female from Liang Toge had an estimated height of 150 centimeters, while two adults from Gua Alo, less than 3,000 years old, were similarly short—the male was 160 centimeters and the female about 148 centimeters. However, individuals from Liang Momer and Liang Panas, closer to 7,000 years old, were much taller. Like earlier hominid populations on the island, Mesolithic hunter-gatherers on Flores seem to have undergone size reduction in response to insular conditions, and in the latter case this took just a few thousand years. It's a well enough known fact, too, that modern people, such as the Andaman Islanders and the Negritos of the Philippines, living on islands for generations, have a small stature, as do the pygmies of the Congo, where small body size is probably an adaptation to their environment's limited resources.

The extinction of *Homo floresiensis* and *Stegodon* on Flores shows again the vulnerability of animals on islands when circumstances change. But what specifically had been the cause of their demise? At Liang Bua, the youngest remains of both species were sealed in by a distinctive white tuff that was produced by a volcanic eruption about 12,000 years ago. If this eruption devastated the island, it could easily have resulted in island-wide extinctions—or so I first thought. But evidence recently pieced together suggests a darker side to the story. Geochemical analysis of the white tuff by Chris Turney and Doug Hobbs now shows that it was not derived from any volcano on Flores, but had been wind-blown from a source about 600 kilometers to the west in the vicinity of Bali. They therefore argued, with the force of the recently converted, that this volcano could not have decimated the whole of Flores; that it was not responsible for the extinction of *Stegodon* and *Homo floresiensis* on the island. Instead of a smoking volcano, we had another smoking gun.

After surviving major catastrophes and periods of environmental change, as volcanoes erupted, ice ages and interglacials came and went, and the island oscillated in size, the only two large-bodied mammals on Flores disappeared just before the first evidence for modern humans at Liang Bua—this in the context of modern human impacts on other pristine island animals, including the extinction of the Australian megafauna around 47,000 years ago, the diminutive cave goat *Myotragus* on Majorca 10,000 years ago, the pygmy elephants and hippos on Cypress 10,000 years ago, the giant elephant bird of Madagascar around 1,000 years ago and moas from New Zealand from 800 years ago. Were we directly responsible for the extinction of the world's last known populations of *Stegodon* and another hominid species, *Homo floresiensis*? On balance of evidence, probably yes. But why was Flores a refuge for so long, and until so recently?

With circumstances very reminiscent of the Balearic Islands in the western Mediterranean, which were quarantined by sea currents from colonization for millions of years, the inaccessibility and refuge status of Flores was also current induced. In fact, if you examine the currents that sweep through the region you'll see that it is well-nigh impossible to swim, drift or raft to Flores from areas of the Asian mainland directly to the west via Java, Lombok and Sumbawa: ripping currents between the islands themselves will simply push you south and out into the Indian Ocean. Elsewhere in the area, currents will carry you west: from Australia and New Guinea back along the Nusa Tenggara chain either toward the Southeast Asian landmass if you are on the northern side of the Nusa Tenggara chain, or out into the Indian Ocean if you are unfortunate enough to be on the south side of the chain.

Not surprisingly, there is increasing evidence that land animals reaching Flores on their own accord came with the currents from the north and the east. The rodents, for instance, have their ancestry with rodent populations in Sulawesi, New Guinea and Australia. Similarly, the few large animals that colonized Flores—giant tortoise, Komodo dragon, *Stegodon* and early hominid—were probably swept there by ocean currents, with Sulawesi and adjacent sections of the Asian mainland being the most likely source areas. Even so, crossing to Flores must always have been difficult and infrequent.

Given "relict" animals on Sulawesi, such as *"Elephas" celebensis*,

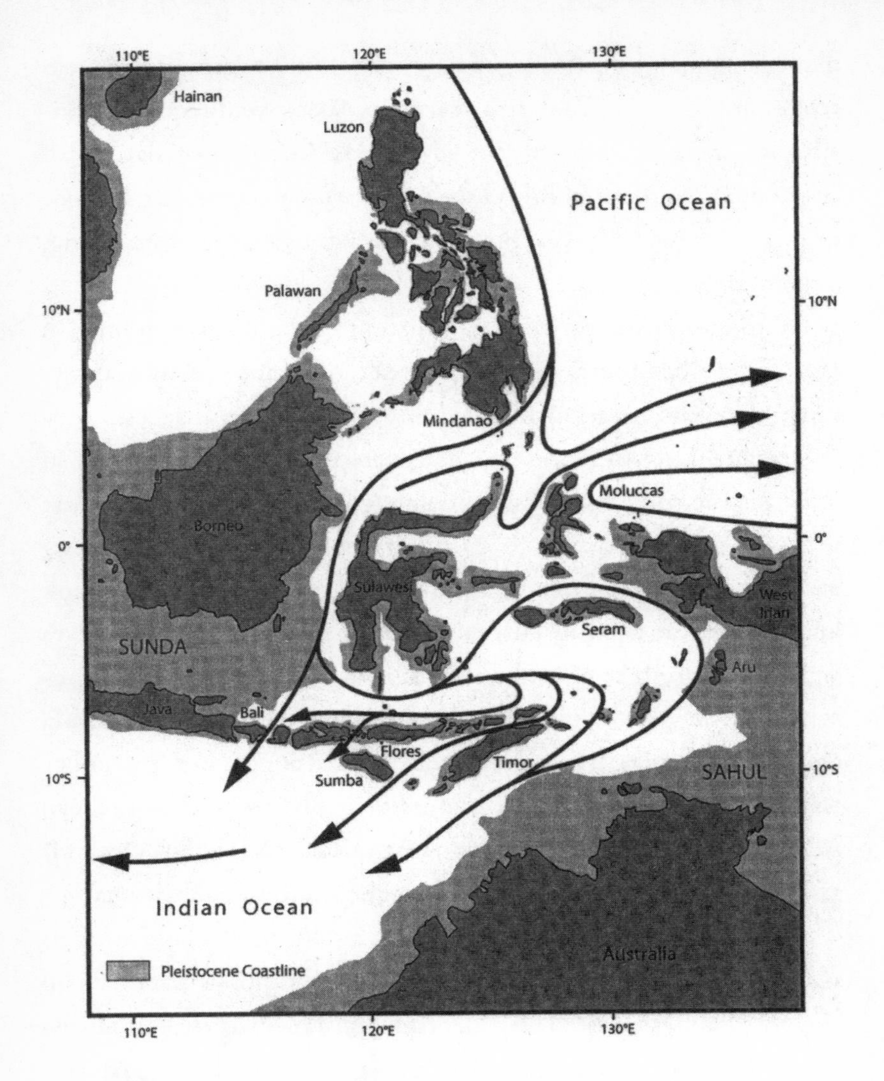

Oceanic currents in island Southeast Asia at times of low sea level, when the sea passage between New Guinea and Australia was closed. The direction of ocean currents and island geography probably both dictated a natural north-to-south pattern of animal dispersal in the region, as also indicated by the interisland distribution of animals in the fossil record. Significantly, modern humans were the only large Asian animals that made the west-to-east crossing to Greater Australia on their own account. (CREDIT: MIKE MORWOOD)

Babirusa, and anoas, which have no known extant or extinct relatives in mainland Southeast Asia but hint at early Indian and African connections, is it so surprising that Flores also served as a refuge—one on which a large monitor lizard, *Stegodon* and an early human lineage survived long after they had vanished elsewhere? *Homo floresiensis* harks back to the small hominids at Dmanisi 7,000 kilometers away, and beyond that to the earliest human species in Africa. Another reminder of how little we know about early hominid evolution and dispersal.

The first modern people to colonize the island also had connections to the north, as shown by the introduction of the Sulawesi warty pig around 8,000 years ago. But the presence of Papuan languages in Timor Alor and formerly Flores strongly suggests that colonists came from the east, more specifically from the Papuan area of Greater Australia. Thus the discovery of *Homo floresiensis* has thrown light on yet another great mystery, that of the colonization of Australia by modern people.

The major currents in island Southeast Asia are determined by the southerly through flow of water from the Pacific to the Indian Ocean, and there is only one area where it may have been possible

A large canoe depicted in a panel of Bradshaw rock paintings in the northwest Kimberley. A number of such paintings are now known. None have sails depicted. (CREDIT: MIKE MORWOOD, AFTER WALSH, 2000)

to make easterly current-assisted crossings to reach Greater Australia—from Sulawesi via the Moluccas and Seram.

Until recently, few would have imagined that the first modern Floresians may have been seafaring Papuan speakers who sailed from the Greater Australian mainland some 12,000 years ago; already practiced root and tree crop cultivation; encountered and extinguished a Lost World fauna, including a distant relative; and by way of compensation imported the pig. But stranger things have happened.

8

The Reaction

While the *Nature* papers were still pending in 2004, Soejono and I organized further work at Liang Bua to continue the excavation of Sector VII to bedrock, find the arms of LB1, and obtain more evidence for the life and times of *Homo floresiensis*. The excavation, again under the supervision of Thomas Sutikna, Wahyu Saptomo and Jatmiko, was scheduled to run from July 10 to September 10, but I had to return to Australia in early August to teach at the University of New England. During my absence, Soejono visited the excavation to check on progress. He only stayed in Ruteng a week, but before leaving called a meeting and, much to everyone's shock, issued written instructions that the excavations were to stop on August 10 "because everyone needed to return to Jakarta to help prepare for an ARKENAS seminar to be held on 23rd August." It effectively dashed our excavation aims for 2004. I knew it all had to do with our objection to the LB1 findings being handed over to Teuku Jacob.

I rushed back to Indonesia—on long-service leave, as the university's vice chancellor insisted that my travel was nonessential—and at Liang Bua found a very depressed excavation team indeed.

They had dug down to a series of superimposed occupation floors with high concentrations of flaking waste, stone tools and animal remains, and had also found more hominid remains, including another complete, tiny lower jaw, a radius and a scapula. Associated animal remains included those of *Stegodon*, Komodo dragons (with cut marks evident on some bones), birds, rats and bats. The hunting of *Stegodon* and Komodo dragons must have been a cooperative activity, presumably involving language. Evidence that *Homo floresiensis* used fire was provided by charred bones and by reddened and fire-cracked rocks—including a cluster of three burned, water-rolled, volcanic pebbles and a circle of five similarly burned pebbles that appears to have been part of a hearth. These were all staggering finds. Stopping work a month early to attend a seminar seemed nonsensical. Fortunately, Tony Djubiantono (ARKENAS director) arrived for a brief visit and saved the day. Seeing the importance of the work, he decreed that we should "continue as long as needed."

Kira Westaway was also back on the scene. She had initiated a program of drilling cores from stalagmites in caves. Like trees, stalagmites contain seasonal growth lines, which vary in thickness depending on the amount of water available. The oxygen and carbon isotope ratios locked into the crystalline structure of calcites in each growth line also provide evidence for the prevalent temperature and the proportions of trees versus grasses in the local environment, respectively. As the age of each growth line can be obtained by U-series dating, and stalagmites can represent tens of thousands of years' growth, Kira's work had enormous value for providing information on the paleoenvironmental context of the site. In fact, by overlapping results from stalagmites of different age ranges, she ended up with a 100,000-year record of local changes in rainfall, temperature and vegetation.

But things never go entirely as planned. When departing Hotel

Sindha to return to Australia, Kira accidentally left one important stalagmite core behind—a rod of calcite about 30 centimeters in length and four centimeters in diameter. Rokus, who is an experienced practitioner of reflexology, salvaged this "discarded" sample and worked hard grinding and polishing it to make a beautifully striated tool for applying pressure to nerve centers on his patients. I was philosophical about it; the sample can be retrieved if needed, and in the meantime it is being put to good use.

The long wait finally ended. On October 28, our scientific papers finally appeared in *Nature*—one year, one month and one week after Peter had begun the analysis of LB1. All this time, the little hominid—holotype of a new human species—had been stored in an old filing cabinet with Thomas holding the only key. The embargo, so important to *Nature*, had held. Very few people outside the research team had seen her, and no one else had photographed or studied her.

Bert Roberts is savvier than most when it comes to public relations and communication. He suggested that we organize a formal news conference to coincide with the *Nature* publication date, at which we would announce the discovery, provide a potted history of the associated research and then field questions. His original proposal was for the conference to be held at the University of Wollongong Business Centre in Sydney, but the publicity officer with the University of New England, Lydia Roberts, was horrified at this and suggested that the Australian Museum was a much more "neutral" venue. While in Jakarta in September, Thomas Sutikna and I also laid the groundwork for an Indonesian press release and conference by ARKENAS to coincide with the planned Australian event. The dates for publication were confirmed and we drafted a list of salient points for Prof. Soejono and Tony Djubiantono to use, but for some reason the Indonesian press release and conference never happened. This has been a point of contention and criticism ever since. The next time we did better. For the paper on the brain of LB1, published in the March 2005 edition of *Science Express*, a press conference was held at the Centre for Archaeology in Jakarta to coincide with the press conference organized by our colleagues in America.

When the discovery was announced all hell broke loose, as the world's media e-mailed and phoned our offices and homes—about 200 inquiries a day for the first week, with Peter doing 100 interviews in the first three days. The interest was overwhelming: we were featured on about 98,000 Web sites and were headlined in about 7,000 newspapers, including the *Guardian, Sydney Morning Herald, Nepalese Times, New Zealand Herald, New York Times*, and a local newspaper in the remote Mayan area of Honduras, which had only featured one story previously in which Australia had been mentioned—the beatification of Mary MacKillop, the first Australian saint.

The story seems to have made it into every major newspaper around the world, into most popular newsmagazines, including in Australia *The Bulletin, Time, Newsweek* and the *Financial Review*, and was reported as news on most TV channels and science programs. "It's always a delight to welcome a new member to the family" was the introduction to the story by one newsreader, while Deborah Smith of the *Sydney Morning Herald* surmised, "The find has also put us firmly on the same evolutionary footing as other creatures on Earth, something that would have pleased Darwin." Comic pieces and cartoons also utilized the information to make social and political comment, with typical examples appearing in the *Australian* and the *Denver Post*.

Political cartoon comparing Hobbit and the Australian prime minister, John Howard. (COURTESY: PETER BROELMAN)

The discovery also generated all sorts of interpretations of philosophical, biological, anthropological and archaeological issues. Prominent anthropologist and author Desmond Morris, presumably an atheist, claimed that "the discovery of a human 'hobbit' on Flores would force many religions to examine their basic beliefs" and that "the existence of Mini-Man should destroy religion."

In contrast, David Wilkinson, a lecturer in theology and science at the University of Durham, was excited because the discovery "poses the big question of what it means to be human." It's a question that has been asked before. A thorough analysis of Neanderthal behavior, for instance, shows that there was little difference between their hunting strategies and those of anatomically modern humans. Neanderthals had long developed sophisticated tools with which they brought down large, dangerous animals, and occasionally they buried their dead and made nonutilitarian "art objects"—at Tata in Hungary, for instance, a sectioned and polished mammoth tooth with adhering red pigment dates to between 116,000 and 78,000 years ago. Strategies used by Neanderthals are also used by modern hunter-gatherers. Indeed, some technologies, such as the development of blades, which are thought to be the signature tunes of modernity, were not taken up by modern people in most areas of Australia, for instance.

Compelling evidence of earlier human species showing modern human abilities comes from one of the most extraordinary archaeological discoveries of recent times—three throwing spears found in Germany, dated at 400,000 years. Because wood usually does not survive in the archaeological record, these spears would be sensational if they were only 3,000 years old. Something a hundred times older is miraculous.

Each spear is skillfully made from the trunk of a 30-year-old spruce tree. This hunter was incredibly powerful. And clever. The tip of the spear is made from the base of the trunk where the wood is hardest. And the center of gravity is a third of the way from the sharp end—just like a modern-day javelin. Only a sophisticated person capable of hunting large mammals in a planned, abstract way could have designed such an innovative tool. But these people were living at least 200,000 years before the accepted appearance of anatomically modern *Homo sapiens*, let alone the advent of their modern behavior.

Paul Davies, professor of natural philosophy and a well-known Australian commentator, said the existence of *Homo floresiensis* "may change our perceptions of ourselves in a similar way to Copernicus' 1530 proclamation that the Earth was not at the centre of the Universe," and that the discovery sidelined humans to "just one twig on what may be forests of life."

Paleoanthropologist Tim White, who worked with Donald Johanson on Lucy, provided a *Star Wars*–type slant:

> This is about discovering a lost world instead of creating it from science fiction. It really existed! [George] Lucas didn't make these creatures up! These are real relatives that nearly lasted to co-exist with us. . . . I think this will become one of the classic examples of human evolution.

Comments by other leaders in the field echoed those sentiments. For instance, Bernard Wood, a paleoanthropologist with George Washington University, said that it was "arguably the most significant discovery concerning our own genus in my lifetime."

While Jared Diamond thought it was the most astonishing discovery in any field of science for over a decade, more surprising to him was the fact that *Homo floresiensis* had coexisted with *Homo sapiens*—a previously unsuspected, tiny species of human living on a remote island in East Indonesia and overlapping considerably in time with us. It was a fabulous, magical story.

Because of the self-reinforcing way that the media works, it's not really possible to disentangle the causes and effects for all this reaction. However, the main reason for such interest was the very nature of the finding—a previously unsuspected, tiny species of human living on a remote island in east Indonesia and overlapping considerably in time with us. It captured people's imagination and had the makings of a good story, in contrast to all the bad stories about death, destruction and degradation that are more commonly reported. The find also shattered some existing assumptions concerning evolution, and showed that thinking about these issues is not just the province of archaeologists or philosophers, but is also of real interest to the general public.

Audiences at public lectures and exhibitions were a direct measure of this interest. The Museum of Natural History in London put on a special exhibition featuring a copy of the LB1 skull and lower jaw, a one-meter-long ruler to represent the height of LB1 and photographs of Liang Bua, the excavations and the finds. In four months it had over 500,000 visitors. At a smaller scale, Peter Brown gave a public lecture at University College London that was packed, while my public lecture at the Armidale Town Hall more than had 300 people in attendance, with standing room only.

There were also direct economic impacts. The Flores Tourist Authority reported that travel to the island jumped by 21 percent, thanks to media coverage of the discovery. Before this, tourism had plummeted because of the economic crisis, political problems in adjacent East Timor, and the Bali bombing. Companies offering travel to Flores reported an upsurge in interest immediately after discovery. For instance, Peter Paka, owner of a Bali-based company that leads tours to Flores, noted a 1,000 percent increase in daily visits to his company's Web site immediately after the discovery—and Flores was the most searched subject on his site.

And rather than quickly peaking, then fading away fast, the reporting of the story actually seemed to gain momentum. Five months later, in March 2005, for instance, 138,000 Web sites had reference to *Homo floresiensis*. But there were also changes in the types of media interest over time—from an initial rush in daily newspapers, to longer review articles in newspapers and magazines, to inquiries from potential authors, literary agents and publishers all wanting to be involved in the writing of a book on the history and importance of the discovery. The angle of reports also changed as the story developed—starting with the excitement of discovery and its implications, and then moving on to debate about the taxonomy of the skeleton and other issues.

Similar changes in coverage occurred in other mass media outlets, and we had many offers from television companies and programs wanting to film our Flores research and the wider context—including from the BBC in Britain (*Horizon*), the ABC in Australia (*Catalyst, Horizon*), CBS in America (*60 Minutes*), Channel 9 in Australia (*60 Minutes, Sunday*), Channel 4 in Britain, the Discovery Channel and PBS's *NOVA*.

In other developments, the American Association for the Advancement of Science voted the discovery the second most important scientific discovery of the year—after the discovery of water on Mars. In Armidale, hometown of the University of New England, the announcement of the discovery was an occasion of great rejoicing. The little provincial university had scored a global first. Our vice chancellor, Professor Ingrid Moses, immediately and unexpectedly offered Peter and me professorial fellowships on behalf of the entire university community. We accepted quickly in case there had been some mistake and she changed her mind.

Other noticed, too. The Australian Federal Minister for Education, Science and Training, the Honorable Brendan Nelson, also sent a letter of congratulation on the discovery, but especially because, "It is one further sign of the extent of cooperative endeavours that Australia and Indonesia can pursue together."

This was a scientific discovery being talked about in villages, towns and urban centers everywhere. It was a topic of conversation in beauty salons and barbershops, school and university staff rooms and every other type of workplace imaginable. Schoolkids in places we had never heard of before were doing class assignments on the discovery and e-mailing us for additional information, while a Los Alamos art teacher, Petr Jandacek, made a ventriloquist's dummy to present educational, entertaining performances on the life and times of the tiny Hobbit-Hominids. This was to complement his many public performances of Otzi, the Tyrolean Freeze-dried Iceman, for the New Mexico Humanities Council.

It wasn't all smooth sailing, though. In preparation for the publications, and preempting the media attention, Bert Roberts had commissioned artist Peter Schouten to paint a portrait of the little lady so that the public could more easily relate to her. Imagine my surprise, however, when I first saw the painting of LB1 a couple of days before the *Nature* publication date. The figure was professionally done and looked very lifelike with a dead giant rat casually draped over one shoulder, but the painting was clearly

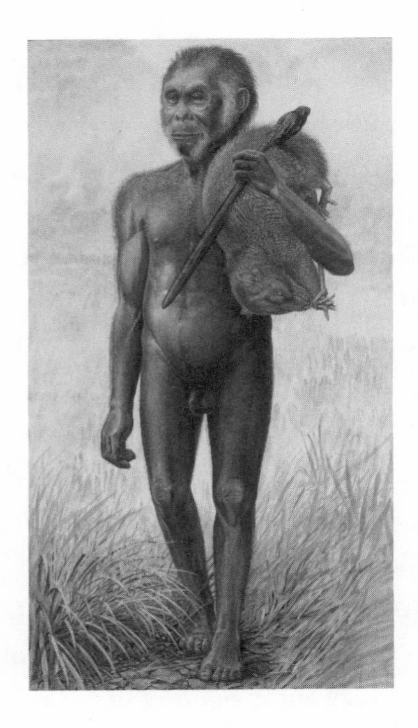

Real-life depiction by Peter Schouten of Hobbit (LB1) with rat, but clearly shown as male! Despite the gender confusion and relatively simian features, this painting has become iconic in the popular press for Hobbit and the associated research. (COURTESY OF *NATIONAL GEOGRAPHIC* AND THE UNIVERSITY OF WOLLONGONG)

male—the penis and testicles were a dead giveaway—while the real LB1 was probably female. Apparently, hunting giant rats was thought to be more of a male activity, hence the sex change. It was too late for me to change it back. Despite these anatomical and gender deficiencies, the painting became such an icon for our Flores research, the discovery of LB1 and associated issues that *National Geographic* and the University of Wollongong purchased the painting for $10,000. One member of the public, however, took umbrage at the anatomical flaw and said she was going to nominate me as Australian Sexist of the Year!

Another major reason for the public uptake of the Flores story was the impact of the Internet. Most newspaper, magazine and journal articles are also posted on the Web, some exclusively, as are the transcripts of TV documentaries and news programs, encyclopedias, and discussion forums—a huge, immediate and very accessible means for getting and communicating information, ideas and opinions. The discovery in Ethiopia of Lucy, a 3.2-million-year-old, partial *Australopithecus afarensis* skeleton, deservedly got great media coverage, for which her nickname helped, but she was found in 1974 in pre-Internet times. Now things would be different.

Publishing in a refereed journal such as *Nature* first was essential to establish credibility, but not too many members of the public read it. To communicate the finding to a much wider audience, including the taxpaying public who were funding our research, we decided to approach a reputable popular magazine with a large circulation. *National Geographic* with a circulation of around 10 million fitted the bill. So just before the start of the 2004 excavations at Liang Bua, I contacted Tim Flannery, high-profile zoologist and author, who was on their Australian Research Committee, to ask whether *National Geographic* would like to cover the story—with the proviso that the embargo be adhered to and with the idea that it could be published ASAP after the papers had appeared in *Nature*.

As usual, Tim must have been very persuasive because later, in Ruteng, I received an e-mail asking if a *National Geographic* photographer, Ken Garrett, could visit the Liang Bua excavation, then in progress. The visit would be in connection with a story on the Flores discoveries planned for the April edition of *National Geographic*. The magazine also wanted us to write a 1,000-word personal account of the discovery of *Homo floresiensis* for the same edition. The authorship fee would be very useful in supporting our research, as would the additional sponsorship money that *National Geographic* offered, on condition that we acknowledge their support in our press announcements and papers. Their April coverage of the Flores research would feature photographs of the excavations and finds, an anatomically accurate depiction of LB1 and scenes from her life based on the archaeological evidence. In addition, National Geographic Television and Film wanted to send a crew to cover aspects of the excavation for a two-hour special they were making on early hominid sites around the world. The answer was yes to all the above, and on request Tony Djubiantono, as director of ARKENAS, faxed Ken Garrett and John Ruben, the producer of the TV special, formal letters of invitation for them to visit and document the Liang Bua excavation.

A mere four days later, Ken Garrett arrived, having flown a horrendous route from San Diego to Bali to Labuanbajo, where he hired a car for the three-hour drive to Ruteng. Ken, a large jovial man, had been with *National Geographic* about 20 years, had been on assignment almost everywhere and was very professional in his dealings with the excavation team and local people. He fitted right in and became part of the team without any disruption, apart from the need for extra lighting in the excavations, and he went about his business using up about 40 rolls of film a day. Three days later the three-man film crew arrived. John Rubin, Dave Linstrom, the cameraman, and Icang, one of the best sound technicians in Southeast Asia, with an Indonesian fixer, Paul Boleh. Bert Roberts, who was handling most of the negotiations, arrived the same day. Film crews are definitely more intrusive than photographers. John had preconceived ideas about what was important, and initially concentrated on the three Australian researchers on site, Bert Roberts, Kira Westaway and me. We had to put a stop to this. It was, after all, an Indonesian site and the excavators Indonesian.

John Rubin and his crew left on August 26, but they must have liked what they saw, because within the week, he e-mailed me again to say that National Geographic Television and Film was now interested in doing an entire film dedicated to the discoveries being made in Flores. After lots of e-mailed negotiations, the second film crew, led by David Hamlin, with cameraman Dave Linstrom and sound technician Icang, arrived in Ruteng on September 30. Part of the deal was that the *Nature* embargo would be adhered to and all people involved would sign confidentiality agreements.

The second film crew was more intrusive than the first and had some funny ideas about how archaeology was conducted. At one stage they had Wahyu down a six-meter-deep trench excavating in the dark with a head lamp for lighting—we had a generator and lighting system in place, but the head lamp scenario was more visual. Another afternoon, they spent over four hours placing colored lights strategically around Liang Bua so that Thomas, Bert and I could be filmed sitting at a small wooden table at night "analyzing" hominid remains in the cave. The visual impact was Gothically spectacular—certainly more so than Room 19 at Hotel

Sindha, which served as our usual bone lab. But what impression did we actually want to convey about the nature of our archaeological work at Liang Bua? In the end it came down to a balance between the demands of our work and David's need to make his program entertaining as well as informative.

We also learned a lot about the making of documentaries—in particular, how a scene has to be filmed over and over again to get the right result. The stock cry during this filming became "*Sempurna. Satu kali lagi*," which loosely translates as "That was perfect. Let's do it one more time." We had one scene, for instance, in which 30 Manggarai workers and the researchers were filmed walking up the road, then entering Liang Bua to begin the day's work. For this a boom crane, flown at great expense from Jakarta along with 600 kilograms of counterweights and associated equipment, was used. It and the associated two-man maintenance crew were costing 6,000,000 rupiah a day to hire—about double the running cost of our entire 40-person excavation team! To capture the whole scene required that the boom be moved in a complex arc as we approached, passed, then entered the cave. This one scene took over two hours and 17 takes to complete, and was never featured in the final version of the film.

David Hamlin also planned to have animation in his film. A very expensive business. His budget would only stretch to animation of LB1 and one animal—either a pygmy *Stegodon* or a giant rat. To set the scene for animated figures in the cave, David was filmed moving in a set path around the site carrying a multicolored ball one meter above the ground. The same path was then mapped out with reference Ping-Pong balls and filmed again. The major problem in doing all this was to control the large crowd of local Manggarai people who clustered around the entrance of the cave to watch all these utterly incomprehensible activities.

Besides the excavations and our findings, there were two other main components to the film. The first was to construct a bamboo raft and paddle from Sumbawa to Komodo Island, which at times of low sea level would have been part of Flores. Sumbawa was then considered the most likely immediate source area for hominid colonization of Flores, and bamboo rafts were a possible means by which this was achieved. This experiment would help us understand some of the difficulties in making and using bamboo rafts, and thereby the degree of intelligence and logistical planning required. Robert Bednarik, an Australian colleague, had already undertaken two such trips—from Bali to Lombok across the Wallace Line, and from Timor to northern Australia. He was the most obvious choice to design the raft, organize its construction and captain the actual crossing. This was another expensive business. Robert was flown in from Australia and five tons of bamboo were purchased in Bali, flown to Labuanbajo, and then shipped across to Bima, at the eastern end of Sumbawa. There with the help of 17 local Sumbawanese, Robert, with superb management and political skills, got the 12-meter-long raft constructed in record time.

It was a 12-person raft, and four of the Liang Bua researchers were to be part of the crew—Thomas, Wahyu, Bert and me. The other paddlers were to be local fishermen. So leaving the backfilling of the Liang Bua excavations to Jatmiko, Rokus, Wasisto and Atiek, we drove in convoy down to Labuanbajo late one afternoon and boarded two fishing boats for the overnight trip to Bima. There next morning we met Robert Bednarik, and inspected the raft being constructed under cover next to a nearby beach. It was indeed an impressive construction, made entirely from traditional materials and with a row of six seats down each side. It was finished that afternoon and after a ceremony was hauled into the water, where it performed perfectly. The raft was then towed to a small nearby island from where we would begin this epic voyage first thing next morning.

Before first light we set off in a convoy of three fishing boats, had breakfast on the island, then set off paddling the raft to Komodo Island, on the horizon about 22 kilometers away. After some initial apprehension on the part of most of us unused to bamboo rafts on the open sea, it became clear that this was a very stable and seaworthy vessel. Rather than bobbing up and down, the construction of long poles lashed together allowed waves to pass through. The major problem was that the raft was heavy, cumbersome to maneuver, and the only means of propulsion was paddling, which meant a slow, tortuous trip. If early hominids were smart enough to make a raft, surely a simple sail made of bamboo strips was not beyond them. A thatched roof to protect the occupants from the sun would also have been a welcome addition. Robert's earlier experiments had shown that such rafts have to be constructed under shelter to prevent cracking of the bamboo in the sun, so why not a shelter on board as well?

It was instructive to watch other small craft in the area. All had sails and shelters, and while we proceeded at a snail's pace, and were consistently pushed south by the strong currents between Sumbawa and Komodo Islands, little outriggers with their colorful sails would appear on one horizon, pass us, and then disappear over the other horizon. It was also obvious that those other seafarers lived on their boats for extended periods. They did not just sail from A to B, disembark and rush inland. The first people to intentionally colonize islands in the region probably used their seagoing craft in the same way. In contrast, we rafters all had other commitments, planes to catch and deadlines to meet. People not only needed the appropriate technologies and planning skills to make deliberate sea crossings, they needed the appropriate mind-set.

The crossing took 11 hours and was much longer than the straight interisland distance of 22 kilometers. We were swept south of Komodo Island by the strong current, then had to paddle back on the lea side of the island to reach the south coast. By then we were all totally exhausted: the trip took 11,000 paddle strokes by my reckoning, and a number of substitutions were made. The first to go was Bert Roberts, who after two hours suddenly began to vomit over the side with seasickness. Because he was one of the central figures in the film, he was replaced by David Hamlin, who donned Bert's shirt and hat for the sake of film continuity. Thomas and Wahyu lasted about six hours before also jumping onto one of the accompanying boats and being replaced by local fishermen. However, all came back on board for the grand finale, the raft making physical contact with the forbidding rocky coast of southern Komodo Island.

As we approached the shore, we could see waves smashing against rocks and no obvious sheltered landing point—potentially very dangerous if we smashed against the coast, or if scraping over submerged rocks cut some of the bindings that held the raft together. The accompanying boats chose that precise moment to assemble for a meeting 50 meters behind us right when we needed reconnaissance and berthing guidance. Whether we should go to the right or left of a particularly nasty-looking large rock slab was not clear. But no one was doing anything. Robert remained seated where he was, farther back, and said nothing. So I yelled, "Let's take it to the left!" which we did. As soon as we touched rock, one of the boats pulled alongside and David Hamlin, Dave Linstrom, Icang, Bert, Thomas and Wahyu leapt aboard to take their places.

By this stage, I was extremely angry that David seemed to place a higher priority on his filming decisions over safety concerns for the raft and those on it. I started to remonstrate with him, and was told that this was the business of the captain, Robert, who then came forward and for the benefit of the cameras began to issue nautical instructions. Feeling totally physically and mentally exhausted, and rather mutinous, I stepped forward to take the matter further, accidentally put my foot between two of the lashed bamboo poles, twisted and fell sideways—almost overboard, in fact. This broke one of the bones in my right foot, so I resumed my seat and calmed down.

The raft trip had been a really demanding rite of passage, and everyone was elated at the success of the project—which a couple of hours before had been in doubt. The film crew souvenired six of the 12 paddles to take back to America, while the raft itself was to be towed back to Bima to take pride of place in a local museum. Our research team and film crew boarded one of the accompanying fishing boats and headed back to Labuanbajo for the flight back to Bali. Within minutes, we were all were sprawled asleep on the deck. Robert had to catch a flight out of Bima the next morning to Bali, then another to Australia, so he boarded the fishing boat towing the raft. However, because of this extra weight, the second boat ran out of fuel halfway and those aboard had to complete the trip on the raft, an all-night trip and with only six paddles. They reached Bima early the next morning beyond exhaustion, soaked to the skin and just in time for Robert to catch his flight with wet passport and travel documents.

The second additional component of David Hamlin's project was to fund and film a comparative study of the brain of LB1. For this he contacted Dean Falk of Florida State University, an acknowledged expert in the field. With colleagues from the

Mallinckrodt Institute of Radiology, she was able to use our CT scan data to make a virtual and then a physical model of the space inside LB1's skull. These models, called endocasts, provide evidence for the shape and structure of the brain that previously occupied the space. Dean wanted to compare LB1 with her broader family: modern human females, a modern human pygmy, a modern human microcephalic, Asian *Homo erectus*, two species of australopithecines, gorillas and chimpanzees. The study was at the cutting edge of research, and expensive—but National Geographic Television and Film would pick up the tab, provided they could include it as a segment in their forthcoming film on the Flores research that they had shot earlier in the year.

We sent Dean a copy of the CT scan data for LB1 on the agreed condition that it would not be used to make other copies of the skull, that it would be deleted from their system after the endocast research, and that our research team would have the chance to input, and would be credited with coauthorship. There were also extreme time constraints on getting any resulting papers published. They had to be published in a serious refereed journal before the National Geographic film was first aired in America. This was a tall order, but the results of the comparative endocast study turned out to be so interesting that the American journal *Science* agreed to get the paper refereed, and to make a decision in record time. Dean and her colleagues were unequivocal. The size of LB1's brain, absolutely and in relation to body size, was the same as that of an australopithecine. Although tiny, the brain had no evidence for pathology, but had the overall shape of *Homo erectus*, as well as a unique combination of primitive and advanced traits. They concluded that LB1 and *Homo erectus* likely shared a common ancestor.

David Hamlin's film was first shown on the National Geographic Channel in America on March 13, 2005, and John Rubin's

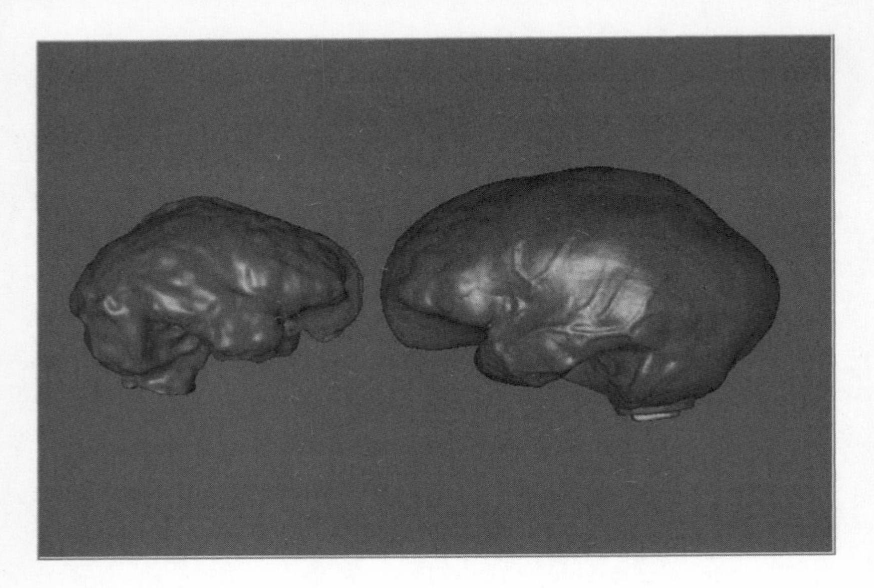

Comparative side views of the endocasts of LB1 Homo floresiensis *and* H. erectus *from Java. Despite differences in size, there are a number of specific similarities. Both are long and low, but LB1 has enlarged frontal lobes, which contrasts with more primitive traits in the posterior section of the brain.* (PHOTO: DEAN FALK)

special the following week. Both went over very well and were on-sold to other networks throughout the world. The film was first shown on the ABC channel in Australia on August 25. Lots of people who don't read much still watch TV, so we were communicating our findings to a very large audience with minimum time lag. The discovery of a new, tiny human species warranted such coverage, which had to be good for the public profile of archaeology.

Despite some hassles, differences of opinion and impacts on research progress, getting involved with *National Geographic* magazine and TV had been well worth the effort. Our research aims, results and, we hoped, enthusiasm were going to be communicated to a much wider audience, we had had considerable input into the nature of the message being conveyed, and other unexpected opportunities had arisen. For instance, our research and results were to be the April cover stories in the American and seminal Indonesian editions of *National Geographic*. The latter, in Bahasa, Indonesia, was also having a press conference in Jakarta to celebrate their launch to which our research team and LB1 were invited. In addition, other rival organizations, such as Discovery and CBS, vied for similar coverage; all of which helped make good media contacts for the future. Being exposed to people with different backgrounds, ideas and angles was also useful. David Hamlin, in particular, had brought fresh aspects of the research to the fore, while the raft trip he organized had been exciting, adventurous and informative—surely all hallmarks of good research.

In a Huxley Memorial Lecture (1958), "Bones of Contention," Wilfrid Le Gros Clark looked back at the history of paleoanthropology to observe that

> [e]very discovery of a fossil relic which appears to throw light on connecting links in man's ancestry always has, and always will, arouse controversy.

I therefore anticipated healthy debate about the identification of Hobbit as a new species, *Homo floresiensis*, and her inferred evolutionary history. At one end of the spectrum, there was John Hawks at the University of Wisconsin, who claimed that LB1 was too different to be included in the genus *Homo*; whereas Susan Anton, at New York University, who has spent many years studying *Homo erectus* fossils in Java, argued LB1 should be just considered as a variant of *Homo erectus*. In contrast, Colin Groves from the Australian

National University in Canberra accepted that the skeleton was of a new human species, but said that she had very primitive traits, indicating that the most likely ancestral species was *Homo habilis*, which preceded *Homo ergaster* in the African evolutionary sequence. But the most persistent criticisms came from Professor Teuku Jacob (Paleoanthropology Lab, University of Gadjah Mada), Dr. Alan Thorne (Australian National University) and Professor Maciej Henneberg (University of Adelaide [Australia]).

Apparently, straight after the papers were published in *Nature*, Thorne and Jacob concurred that LB1 was a microcephalic modern human, while Henneberg published an article in the *Adelaide (Australia) Sunday Times* to the same effect. Then Thorne and Henneberg got together and wrote an article for the on-line journal *Before Farming* to which Peter Brown and I got a chance to respond. Peter began his piece by throwing down a gauntlet: "The authors have either not read the article upon which they are commenting, or have a very limited knowledge of hominin anatomy, perhaps both," and that no substantial peer-reviewed journal would have published their paper. He went on to explain that many of LB1's traits had nothing to do with microcephalia—the primitive roots of the teeth and the shape of the pelvis being just two. Thorne and Henneberg were apparently outraged. They considered the response unethical. It probably didn't help that their papers submitted to *Science* and *Nature* were rejected. But they were still to have their day in court—or at least in the subsequent trial by media.

Genetic disorders can result in stunted growth, as well as microcephalia, and the smallest, well-documented height for a modern human adult is 90 centimeters. But to suggest that a population of microcephalic, modern human pygmies, about a meter tall and with brains about one-third the size of normal people, lived at Liang Bua for more than 80,000 years is stretching credibility a bit.

Thorne and Henneberg were suggesting that modern people had arrived on Flores by 95,000 years ago—a sensational find in anyone's language—but that they were all retarded!

The comparative study of LB1's brain endocast by Dean Falk and colleagues showed that it *least* resembled that of a microcephalic! This conclusion was borne out in a second study of the endocast by a rival group led by Ralph Holloway, from Columbia University in New York, another acknowledged expert in the field. But there was still a problem. If *Homo floresiensis* were simply a scaled-down version of the larger-bodied *Homo erectus*, then the brain should be bigger. One of our critics, Professor Robert Martin, a primatologist with Chicago's Field Museum of Natural History, first pointed this out. He noted that there is a fundamental law in biology that says that if body size is halved, then brain size reduces by only 15 percent. Disingenuously, in making this claim and formulating possible models for the dwarfing of LB1, he chose not to include detailed data on the evolutionary history of the cave goat *Myotragus*, "because the mainland ancestor is unknown." True enough, but there is certainly enough data to show that the relative reduction in size of body and brain of *Myotragus*, during its five-million-year isolation on the island of Majorca, does not fit his "fundamental law."

Assuming that the mainland ancestor of LB1 was *Homo erectus* with a stature of about 175 centimeters and a brain of 900 cubic centimeters, he estimated that the half-sized hominids at Liang Bua should have had brains of about 750 cubic centimeters. The conclusion reached by Martin and others was, therefore, that LB1 was not normal, but was a microcephalic modern human pygmy. To explain why the remains of at least 13 dwarfed, chinless, long-armed individuals with dental and other postcranial "anomalies" were deposited at the site between 95,000 and 12,000 years ago, they further suggested that the local population was subject long-term to a group of inherited genetic disorders, and that

> [i]n addition to genetic factors increasing the likelihood of microcephalics occurring together, it is conceivable that cultural factors might have enhanced this, as at a recent religious site to which microcephalics were brought.

Why the reputable journal *Science* agreed to publish this nonsense remains a mystery.

Martin was perhaps reacting to the proposition that a recent human species had retrograded from the onward-and-upward progress long thought to have characterized hominid evolution, but of course that is exactly what an island dwarfed mammal might be expected to do, human or no. In such cases, the biological "law of economy of growth," first espoused by Charles Darwin, seems much more fundamental. Darwin was actually discussing the reduction in brain size that accompanies animal domestication. The brains of domestic pigs, for instance, are about 34 percent smaller than those of their wild counterparts, without any comparable reduction in body size, presumably because they do not require "vigilance" to avoid predators. But his general point was that reduction in superfluous parts, especially in brain tissue that is energy expensive to develop and maintain, "will be seized on by natural selection, for it will profit the individual not to have its nutriments wasted in building up a useless structure."

Hominid brains are particularly expensive to build and maintain. In modern humans for instance, the brain comprises about 2 percent of body mass, but consumes about 20 percent of our energy requirements. Hominids appear to have been isolated on Flores—an unstable, resource-challenged island on which a reptile was the only large predator—for a minimum of 830,000 years. It is hardly heresy to suggest that they would have been subject to the same evolutionary pressures to downsize body and brain size, as evident in other large mammals on islands throughout the world.

Martin also assumed that *Homo floresiensis* was a descendant of *Homo erectus*. But, as we have argued, there is a more straightforward, even parsimonious, explanation. LB1's brain size, quantitatively and relative to body size, more closely approximates that of the much older australopithecines—as does her stature and body proportions. It is more likely that *Homo floresiensis* actually evolved from an early small-bodied, small-brained *Homo* species. The finds at Dmanisi with their confusing mix of size and attributes show just how little we actually know about the earliest members of genus *Homo*—both in Africa and Asia.

At some point, assumptions cease to be scientific and instead become ideological. This was not new. In fact, I felt like I'd been thrown back to the century before last, into an anachronistic moment of déjà vu. There is a long history of misinterpreting unexpected finds of fossil hominids in terms of pathology, stretching back to 1857, when the first known Neanderthal skeletal remains were found in the Neander Valley, Germany, during quarrying of limestone. Comprising the top of a cranium and some leg and arm bones, they were dismissed by one expert as a Mongolian Cossack from Napoleon's retreating army, with the bowed legs a result of a life on horseback. Others said that the abnormally shaped skull was probably from one of the "savage" races, mentioned by Roman writers, living in the northwest of Europe before the Celts and Teutons, or attributed the bones to a still-extant race, either an old Dutchman, a Celt, or indeed just an idiot.

A more generally accepted interpretation was that the bowed

legs were the result of rickets caused by a vitamin D deficiency. Pain from the disease had also caused the individual to habitually furrow his brow—hence the prominent brow ridges. The preeminent pathologist of the day, the German Rudolph Virchow, agreed. In 1870, he pronounced that the Neander remains were modern and pathological—the individual had suffered from rickets. Also, the advanced age of the individual indicated that he had lived in a relatively recent sedentary society. Virchow, self-described as "the Pope of German Science," was a long-standing opponent of the theory of evolution and his misdiagnosis of the anatomical evidence was undoubtedly colored by his ideological position.

Similarly, in 1891, when Eugène Dubois announced the discovery of Java Man, the type specimen of *Homo erectus*, the British paleontologist Richard Lydekker, an ideological opponent of Darwinian evolution, wrote a review in *Nature* in which he dismissed the skull as that of "a microcephalic idiot, of an unusually elongated type." The 74-year-old Virchow also derided Dubois' work, saying he would make short work of his ideas: in this case, he concluded that the skullcap was not pathological, but from some giant form of gibbon and clearly did not belong with the associated fossil femur that was from a modern human. Virchow's view was not modified in the light of the endocast that Dubois made from the inside of the skullcap, indicating that the brain structure was more human than apelike. More than 30 years later, the discovery by Raymond Dart in South Africa of the Taung Child, the first known australopithecine fossil, and the type specimen for the genus, met a similar dismissive response by some senior people in the field. Following Dart's 1925 paper in *Nature* describing *Australopithecus africanus*, or "Man-Ape of South Africa," Sir Arthur Keith, a preeminent English paleoanthropologist, concluded that "[a]t most it represents a genus in the chimpanzee or gorilla group."

As history shows, there is a downside to scientific research, which by its nature involves a balanced tension between differing paradigms, between continuity and change, and between established and upcoming. Skepticism and rigor in assessing new findings and claims are fundamental in science, but so are objectivity, an open mind and the capacity to take on board the unexpected. The cut and thrust of intellectual debate and controversy can quickly become personal, particularly when the stakes are high and there are entrenched positions to defend. The saga of *Homo floresiensis* was about to take another turn.

9

Something Old, Something New, Something Borrowed, Something Blue

The news that some of the Liang Bua hominid remains had been borrowed came like a bombshell. On November 1, 2004, while I was still in Australia, Soejono had prepared a letter for Jacob authorizing transfer of all the Liang Bua hominid remains from ARKENAS in Jakarta to the University of Gadjah Mada (UGM) palaeoanthropology lab in Jogyakarta "for cleaning, conservation and analysis." The letter was on Soejono's personal letterhead and done without the knowledge or authority of the director of ARKENAS, Dr. Tony Djubiantono, who was later informed in a similar letter that Dr. Mike Morwood, as the University of New England (UNE) institutional counterpart, had been consulted and agreed. This was not correct.

Soejono had organized a meeting beforehand with Thomas Sutikna, Jatmiko and Wahyu to state his intention of handing the material over to Jacob unconditionally. They objected strenuously.

Thomas Sutikna and Wahyu Saptomo analyzing Liang Bua hominid remains at ARKENAS in late November 2004, just before the remains were taken by Teuku Jacob to the UGM palaeoanthropology lab at Gadjah Mada University, Jogyakarta. (PHOTO: MIKE MORWOOD)

But Thomas, under enormous pressure, finally surrendered the key to the locked filing cabinet containing the Liang Bua hominid finds. Jacob duly arrived on November 2, and with another letter on UGM letterhead, to no one in particular, he took responsibility for the fossils and thanked Soejono. The deal was done and the skull, lower jaw and femur of LB1, plus the new lower jaw that we had not yet had time to study, left for Jogyakarta. Since no record was kept of the items taken, we worked out later what was missing by a process of elimination.

On November 23, I flew to Jakarta to confirm plans for the next phase of fieldwork, and to find out what had been removed to

Jogyakarta. By this stage, Jacob was going public about the skeleton being that of a dwarfed, modern human with microcephalia, and on the basis of the roundness of the eye sockets rather than the more gender-specific pelvis had reinterpreted the 30-year-old adult female as a male. He also took the opportunity to say that none of our research team would be involved in future analyses of the material. Claims that it was a modern human microcephalic only about 7,000 years old helped obscure the fact that he had taken hugely important finds and the associated intellectual property from another Indonesian institution and research team.

When Soejono and I first met after these events, he was evasive. He did not want me discussing anything with Tony Djubiantono, and said that Jacob was coming back for the remainder of the material on December 1. His attitude to the younger Indonesia colleagues was that "those boys" should accept his decision as institutional counterpart. It was not their business or place to complain. Those "boys" were, in fact, all men in their 40s or 50s with years of archaeological experience.

Taking Jacob's avowed aim to return in December for the remaining Liang Bua finds seriously, Thomas, Jatmiko, Wahyu, Rokus and I photographed and measured the finds on the assumption that we might never see them again. The one really troublesome absence was the new lower jaw, proof positive that LB1 was not the only tiny human at the site. I wrote a formal letter to Tony reemphasizing the terms of the agreement between my university and ARKENAS, and that the removal of the Liang Bua finds was a flagrant breach of them. I asked if Tony would send a letter to Jacob requesting immediate return of the missing material, and stating that future access to material in the ARKENAS collections depended on his compliance. Since this was an issue between institutions, a copy of the letter also should have been sent to the rector of the University of Gadjah Mada.

Raden Pandji Soejono and Teuku Jacob conferring at ARKENAS on December 1, 2004, just before Jacob took the remaining Liang Bua hominid finds back to the UGM palaeoanthropology lab. (PHOTO: MIKE MORWOOD)

There was no response from Tony, but Soejono was busy making preparations. He wrote a letter for me to sign that said that I agreed to Jacob taking the material. Having already discussed this with other members of the research team, who unanimously opposed removal of the remains, I refused. Next day, Jacob arrived with an assistant, Koeshardjono, and met with Tony to request permission to borrow the remaining finds. Tony called a staff meeting immediately to discuss the situation and possible courses of action, while literally in the next room Soejono, Jacob and Koeshardjono proceeded to systematically empty the remaining Liang Bua hominid finds from the filing cabinet and stow them in a brown leather case brought specially for the purpose. The meeting was still in progress as Jacob and his assistant, case in hand and looking decidedly uneasy, descended the stairs, left the building, got into a taxi and sped away. I gate-crashed the meeting and asked Tony if he was aware that Jacob and the bones had just left. Soejono confirmed this news and immediately the meeting broke up in noisy disorder.

Jacob and Koeshardjono were summoned back, case still in hand, to have lunch and further talks with Tony. We were now all confident that the plot had been foiled, and that at the very least Jacob would have to study the bones at ARKENAS, not take them back to Jogyakarta. So imagine our shock when Tony came in saying that "to let things cool down and because Jacob is an old man," he had let him take the remains back to Jogyakarta for a month—until January 1, 2005. ARKENAS even ended up paying Jacob's travel costs!

That night I was sick to the core—felt like my chest was going to explode. Probably a delayed reaction to disappointment, frustration and anger. We had been let down again. Was much more positive in the morning. We needed to get publications on the new finds out as soon as possible in quality scientific journals, to "claim" them and the credit, and to make sure that the evidence and our interpretations were peer-reviewed and available for scrutiny. First and foremost, all the contextual information on the finds—the stratigraphic sections, plans of associated occupation floors, stone artifacts, faunal remains and charcoal samples—had to be put in a secure place. Jacob had all the *Homo floresiensis* remains but no information on the provenance of the 2004 finds. For all he knew, they might just as well have come from the surface of the moon.

We did not have to wait long before Jacob put out a flurry of public announcements and press releases. There were articles in two Indonesian newspapers, *Kompass* and the *Jakarta Post*, showing him posing proudly with the trophied LB1 skull. In an article that was in a British newspaper, the *Guardian,* he not only restated that LB1 was a modern human male with microcephalia, but that there was now a battle raging between him and the Australian researchers, who could not tell Indonesian scientists what to do; could not play the role of "sheriff"—a reference to Australia's role in East Timor. Ironically, Jacob's rationale for borrowing the hominid remains, and playing the nationalist card, was greatly resented by younger Indonesian researchers at ARKENAS, the University of Indonesia, Balai Arkeologi Jogyakarta, University of Gadjah Mada and many other institutions.

Jacob had not had so much attention in years and seemed to be enjoying the limelight. But why these claims were blithely reported

Teuku Jacob with trophied LB1 skull in 2004. (photo: harun yahya international)

verbatim in reputable newspapers is a matter of conjecture. In the case of the *Guardian*, all the reporter had to do was wander down the road to London's Natural History Museum and talk with someone like Professor Chris Stringer to put Jacob's proclamations in perspective. Such media reports can automatically distort scientific debate by giving "equal time" to both sides without regard to actual evidence. This was all grist for the mill for fundamentalist Christian and Islamic voices, which chimed Jacob's claims that LB1 was a modern male microcephalic. We had got it all wrong, and were bending and twisting the evidence to argue our case, when the truth was that evolution itself was based on lies and misinformation.

The January 1 agreed deadline for Jacob's return of the Liang Bua hominid bones came and went, and I despaired that the remains would ever be returned. Meanwhile, other aspects of the research continued. Excavations at Song Gupuh, a large limestone rock-shelter in eastern Java, were due to begin in early February 2005, and while in Jakarta making preparations, I had a chance to talk with Tony Djubiantono, Soejono and other senior researchers at the ARKENAS. By this time, Tony was feeling very uncomfortable with continued questions from the Indonesian and foreign press about the fate of the bones, and in a fit of optimism had purchased a large safe with a combination lock to house them when returned. It stood lonely, forlorn and empty in his outer office. I suggested that he could keep his beer cold in it.

Soejono was also looking very stressed. He had been sidelined, was being largely ignored by many other researchers at ARKENAS, and had been replaced as their institutional counterpart by Tony in a new Agreement of Co-operation that Tony and I now negotiated and signed. As a gesture of conciliation, I asked Soejono whether he would be willing to visit the Song Gupuh excavation. He replied that the other researchers would not welcome him there. "Nonsense, Pak. We would all value your opinion and advice on aspects of the work there." He seemed relieved, and an advance of 3,000,000 rupiah (about U.S.$330) was then handed over so he and his colleague Ibu Bintarti could fly from Jakarta to Jogyakarta and the nearby excavations for a few days. He nominated February 20 as the day he would arrive at Song Gupuh.

On February 9, while the Song Gupuh excavation was in progress, Carina Dennis, a reporter with *Nature* based in Australia, phoned us at Punung. She had heard that Professor Jean-Jacques Hublin, a researcher from the Max Planck Institute for Evolutionary Anthropology in Germany, had taken some of the LB1 bone from Indonesia back to Germany for DNA analysis. Rex Dalton, an American reporter with *Nature* who was with us at the time, immediately spoke to Carina and got the details, then rang Hublin, who admitted taking two LB1 bone samples to Germany, but claimed that he had the permission of Jacob, and if there were problems with this, it was not his concern. With further probing, Hublin became more evasive, saying he could not remember when he'd come to Indonesia to collect the samples and anyway that "no analyses would be done until the matter was clarified." A researcher from the Max Planck Institute—one of the world's most prestigious scientific organizations—seemed to have been involved in removal of important hominid remains from Indonesia without proper authorization. Rex seemed very pleased with his evening's work.

The next day, Tony Djubiantono arrived in Punung to do interviews with a CBS team from the American *60 Minutes* program. They had just done a tour of Flores with Doug Hobbs, paying visits to Liang Bua, to Boawae to interview locals about the Ebu Gogo stories, and to Rinca Island to see Komodo dragons. Tony knew nothing of the DNA samples taken from the skeleton and whisked away to Germany, despite the fact that LB1 was officially part of his institution's collection and was only on loan.

The following morning, we had some more upsetting news for Tony. After their Jacob interview, the CBS film team told us that *60 Minutes* from Australia's Channel 9 was sending a film crew to interview Jacob, but had also paid for Alan Thorne, our critic at the Australian National University, to accompany the crew from Australia and were planning to film analysis of the remains. For good measure Maciej Henneberg, the paleoanatomist from Adelaide who had claimed that LB1 was a modern human microcephalic, would join them. *60 Minutes* wanted to interview us afterward, and take a tour of Flores with Doug Hobbs. In previous years we had approached various media people to ask if there was a story in our Indonesian research—without success. Now one film team after another was lining up to come and do the story.

Of course our first inclination was to tell *60 Minutes* where to stick it, but Rex Dalton recommended otherwise. It was important, he said, to take the opportunity to present our story on the program concerning the taxonomic status of LB1, and the fact that two other Australian researchers with no connection with our project still felt free to waltz in and help themselves to our findings at the behest of Jacob, who had "borrowed" the material. We met with Nick Greenaway, the producer of *60 Minutes*, over breakfast. We would cooperate with their filming, but it was important to provide some of the background before Nick and his team met Jacob the next day. Nick was relatively young—mid-30s, maybe—and seemed keen to use any background information that would enable him to include "hard questions" in his coming interview with Jacob, for a more balanced program.

Other media people were also making contact. Dagmar Roehrlich, a Germany journalist, was particularly interested in the Max Planck Institute saga—a sad story she called it, and not a good reflection on German science—while a crew from *Sunday*, another program from Australia's Channel 9, asked whether they could come to Punung for interviews with us after interviewing Jacob, Thorne and Henneberg on February 21, to be followed by the now standard Flores tour. Jacob used their proposed schedule to try and further push back the date for the return of the remains. After an angry phone call from Tony Djubiantono, who by this time recognized that continued acquiescence was making ARKENAS look weak, Jacob agreed to return them on February 23. In an attempt to steer the public relations and political agenda rather than be steered by it, we then made arrangements with the *Sunday* team to film the hand-back.

The next day at Song Gupuh, we were very curious to find out from the *60 Minutes* crew what conclusions Jacob and Thorne had reached after their analyses. Surely in the face of the many primitive traits of LB1, and the range of additional remains excavated in 2004, including another tiny adult lower jaw, they couldn't still conclude that LB1 was a microcephalic modern human. But that is exactly what they did. Thorne even said that we "seemed to lack the capacity to recognise a village idiot when they see one"! The Liang Bua population, they concluded, was prone to microcephalia, and anyway they were not as small as we had estimated. Microcephalia could be an inherited recessive gene, and the whole Liang Bua population, they said, had suffered from this condition. Their analysis had been done without any information on their context or age of the bones, but as Thorne told my colleague Doug Hobbs later at Jogyakarta airport, he was "not concerned at all with the context. Only with the skull."

A couple of days later a brief article in the *Jakarta Post* confirmed their diagnosis. It said that a team comprising three Indonesian researchers—Professor Dr. Teuku Jacob (head of the paleoanthropology lab and former rector of the University of Gadjah Mada), Professor Dr. R. P. Soejono (head of the National Archaeology Research Centre) and Dr. Etty Indriati (Jacob's assistant)—with three foreign researchers—Professor Dr. Alan Thorne (Australian National University), Professor Dr. Maciej Henneberg (University of South Australia) and Professor R. B. Eckhardt (Pennsylvania State University)—had studied the finds. They had concluded that LB1 had the pathological condition of microcephalia, and was therefore not a new human species, as claimed by Mike Morwood and Peter Brown (University of New England). In terms of numbers, titles and credentials we were apparently outgunned.

Kompass carried basically the same article as the *Jakarta Post*, but also quoted Thorne, saying that Hobbit was probably associated with farming and animal domestication! Since he had little contextual information, where he pulled that one from can only be left to the imagination. We had, they said, clearly been too hasty in our rush to analyze and publish the find as a new species—this from a group that had their analysis of the material and pontifications televized, and their findings "published," with no supporting evidence or peer review, in two national newspapers on the days immediately following their study. To add salt to the wound, Soejono had used the money we had given him for the flight to Jogyakarta to appear in the Jacob media event rather than visit the Song Gupuh excavations.

This charade was actually far more annoying than Jacob borrowing the finds and reneging on promises to return them. In particular, the involvement of fellow Australian researchers, with an

American we'd never heard of before, was something else again. They had used the opportunity to come in and analyze the Liang Bua hominid remains, including finds made in 2004 that we had not had a chance to study. Jacob had invited the media to film their analysis and our unpublished material, and had promptly put out unsubstantiated claims in a press release. No supporting evidence was given, and some of the statements were misleading. Nor was there any peer review of the evidence or claims. And the public display and discussion of the new finds was a breach of the embargo conditions for papers that we had under review with serious scientific journals. There was every prospect now that our papers would be rejected because of this breach of embargo.

Thorne, Henneberg and Jacob were all of the "multiregionalist school" of human evolution, which holds that modern peoples developed in different parts of the Old World from earlier nonmodern populations—Australian Aborigines from Indonesian *Homo erectus* as evident in the extremely robust skulls from Pleistocene Australian sites, such as Kow Swamp and Coobal Creek; the Chinese from Chinese *Homo erectus*; and so on. Some have argued that the identification of a new human species on Flores undermines the "multiregionalist" position. Ideology may explain why Jacob and his cohorts were so adamant that Hobbit was a modern human, and this in the face of a mounting body of evidence that said otherwise.

My interview with Charles Woolly from the Australian *60 Minutes* program took place with these thoughts in mind. This was not a scientific debate between two groups of scientists carefully weighing up the evidence in a learned and objective manner. This was about other researchers borrowing the results of our hard work, as generously funded by the Australian taxpayers via a grant from the Australian Research Council. This was about others undermining a long-term and productive cooperation between Australian and Indonesian institutions and impeding our ability to analyze, publish and get proper credit.

The *Sunday* crew then visited us at Song Gupuh, having just conducted their interview with Jacob. Their strong impression was that he would not be handing the remains back on February 23, or at any other time. Jacob's take now was, "The Centre and the paleoanthropology lab had strong links and were basically united, so why the need?" There were clear precedents. For instance, the Ngandong and Mojokerto hominid fossils were originally excavated by staff of the Geological Survey of the Dutch East Indies, the predecessor of the GRDC, but when repatriated to Indonesia from the Netherlands in 1972 by von Koenigswald, who was Jacob's mentor, the fossils went to the UGM paleoanthropology lab. Similarly, when hominid fossils from the Institute of Technology Bandung were repatriated to Indonesia in 1993 after the unexpected death of Professor Sartono, who had accompanied the collection to the Netherlands for a conference, they also ended up in the paleoanthropology lab collections.

Meanwhile, the Song Gupuh excavation was going well despite the fact that we encountered massive blocks of rockfall in the excavations. We first employed contractors with heavy drilling equipment to help break up some of the slabs, then used sledgehammers, chisels and crowbars. The sheer size of the rockfall slabs had been daunting, so we had had to come up with a more explicit strategy of "limited objectives." The clearing would be done in achievable stages, the first step being to open up the minimum size of trench required to excavate the deeper deposits, with the option of increasing the size of the excavation area if the extra work was justified by findings. The workmen had already intuitively adopted such a policy: large blocks were broken up in workable-sized chunks from the edges. Attempting to smash whole blocks directly was a waste of time and energy. Should the same principle apply to our research and publication strategies? Almost certainly. Do things in bite-sized chunks, but underneath it all keep a strong sense of purpose and direction, in this case defined by the basic aims of our project. Maybe the same principle applies to life generally.

The assigned day of the hand-back arrived. The Song Gupuh excavation team drove to Jogyakarta, accompanied by the *Sunday* film crew. On the same day, Rokus flew into Jogyakarta with a letter from Tony Djubiantono to Harry Widianto, a paleoanthropologist and also head of Balai Arkeologi Jogyakarta, the regional office of ARKENAS. The letter gave Harry the authority to get the remains back from Jacob. After picking up Rokus from the airport, Harry, Atiek (our representative from the Song Gupuh excavation), Doug Hobbs and the *Sunday* team went straight to the UGM paleoanthropology lab for the hand-back.

Jacob appeared friendly and smiled a lot. He confided to Doug that he had been pressured "from above" to return our finds, and that Soejono, who was finding his hostile treatment at ARKENAS intolerable, now wanted the bones back where they belonged. The bones were carefully wrapped and packed in Rokus's suitcase. The only minor hitch was that Jacob wanted to retain three of the bones of LB1, the femurs and a tibia, for "further study." Harry Widianto agreed, provided a date was set for their return also. We subsequently discovered from a journalist contact that Henneberg thought that the leg bones of LB1 were also pathological, and she surmised that this might have been the reason that the femurs and tibia had been retained.

With most of the Liang Bua finds safely in hand, Harry Widianto and company returned to the offices of Balai Arkeologi Jogyakarta to be met by a throng of happy people on the steps outside, including Thomas, Jatmiko, Wahyu, Wasisto and staff from Balai Arkeologi, plus the *Sunday* film crew that had arrived earlier just to record the moment. In one of the offices, each bone was carefully unwrapped and examined for wear and tear. When all the specimens were laid out on a table together, it was like was seeing them for the first time. They were spectacular! Quite breathtaking and so unusual, especially in size. There were the skull and lower jaw of LB1, like old acquaintances; plus the second lower jaw, smaller than the first; a distorted radius, from another individual, which had broken in life and healed badly around 15,000 years ago; and a minuscule radius from a child, who died about 12,000 years ago.

Assisted by Harry and a borrowed pair of callipers, I took measurements on the second lower jaw. Referees on our submitted paper describing the 2004 finds had requested this information, but we had not been able to comply until now. But there was something wrong with the shape of this lower jaw. When compared with photos that Thomas Sutikna had taken previously in Jakarta, the dental arch and the overall width were much narrower. The bones had been doctored!

A closer look showed that molds had been made of the LB1 skull and both lower jaws, using at least two types of latex rubber, of which there were pieces still stuck to the soft bones. Molding should never have been attempted on such fragile remains—that's why we used the CT scan data to make copies of the LB1 skull and lower jaw. On studying the bones closely, we could see that, contrary to usual practice when making molds, the remains had not been coated with a release agent beforehand. Instead, the latex had been directly applied to the bones and had soaked into the porous surface. When removed, the mold had pulled away parts of the outer layer of bone from the skull. The left cheekbone and two teeth had also broken off, and had been glued back. Worse, much of the finer anatomical detail on the base of the skull had been ripped away, probably embedded in the latex mold. Long, deep cut marks were etched along the bottom edge of the LB1 lower jaw on each side, where the latex had been cut away with a sharp instrument. An attempt had been made to conceal the cuts by infilling them with some substance. It is ironic that 3-D resin replicas of the LB1 skull and mandible, made using CT scans we obtained at Fatmawati Hospital by mistake, now provide the most accurate record of the finds as excavated.

The damage to the new lower jaw was even more shocking. When the latex had been pulled away, it had shattered at the symphysis (the front of the lower jaw where the two halves meet), with loss of bone and destruction of the internal, shelflike buttress—one of the most primitive anatomical traits of this new hominid species. The lower jaw had been hastily and roughly reassembled, was misaligned, and was about 20 percent narrower than it had been. Smaller fragments were just shoved in and concealed in a thick coating of glue. Harry and I stared at each other in dismay. It was totally irresponsible, destructive in the extreme, and the antithesis of ethical scientific investigation—sickening, in fact. The damage was irreparable. Molds had been made, and had continued to be made, regardless of consequences.

In addition, other important finds had been damaged, probably in transit to Jogyakarta. For instance, a very fragile and distinctive part of the pelvis, which had archaic australopithecine-like traits, had been smashed. This had provided evidence on body size and shape, muscle locations and how LB1 had walked—but was no more. Other damage was much more difficult to explain. For instance, her tibia had been rebroken along previous breakage lines and glued back together, all twisted out of alignment. Some of the anatomical detail had been lost and lumps of glue were still adhering. The little lady from Liang Bua had been desecrated. I had a strange, sad, disorientated feeling that she should have remained undisturbed in the fine clays of a remote cave on a forgotten island, where she had lain safely for 18,000 years.

That afternoon, the remains of Hobbit and her kin accompanied Rokus on the flight back to Jakarta, where they were immediately taken to the offices of ARKENAS and locked into the safe. The next day, Rex Dalton e-mailed me to say that Tony Djubiantono had just given permission to Jean-Jacque Hublin at the Max Planck Institute to carry out DNA analyses on the two samples of bone taken without proper authority to Germany. This had obviously upset Rex, who was writing an article for *Nature* on the ethics and legality of this sampling. The German media had also been pursuing the story and there was no telling where all this pressure would eventually lead.

The explanation for this was that Hublin, and therefore Jacob, were coming under increasing press scrutiny about the exported bone samples, especially in Germany. So on February 22, Jacob sent Tony a fax on University of Gadjah Mada letterhead asking for him to give permission retroactively. Jacob also provided the wording for a brief letter for Tony to sign and forward to Hublin:

> Dear Dr J-J Hublin,
> The Archaeology Section authorized you to study the Liang Bua samples regarding aDNA testing, collagen dating and other chemical analyses. Thank you for your co-operation.

Tony agreed, in his words, "to try and cool things down." Now Hublin and his samples were legitimate and he could proceed with his analyses.

It was crucial to put the damage to the Liang Bua hominid remains, and the circumstances, on public record immediately. Any delay and Jacob would definitely disclaim responsibility and say that the damage had been subsequently inflicted on the material while in the custody of ARKENAS. I met with Tony Djubiantono to discuss the need for sending a letter to Jacob briefly describing the damage to the major items, noting that the making of molds in the paleoanthropology lab seemed to be the main cause of the damage, and requesting that a copy of the skull and two lower jaws from the molds be provided to ARKENAS. He concurred with all the points, but said he had had a major row with Jacob over the telephone the previous day about the damage to the Liang Bua hominid bones, and to send the letter would be to repeat the points he had already made. He "would now rather concentrate on planning future collaborative research."

Stalemate, but things were moving on other fronts. Journalists from *Science*, *Nature*, *Discovery*, *National Geographic* and many newspapers had got wind of the story about the hominid bones being damaged, and began e-mailing and telephoning Tony Djubiantono, Jacob, Harry Widianto, Thomas Sutikna and me for more information about the damage and its cause. Jacob predictably disclaimed any knowledge of the damage and specifically denied making molds of the skull and lower jaws, despite extensive evidence to the contrary, including the statements of people who had visited Jacob's lab and had seen three of his assistants working on the Liang Bua hominid remains.

As the pressure mounted, Jacob began to refuse to speak with journalists—even hung up on Rex Dalton, who was doing a review story on the damage for *Nature*. By way of compensation, Rex then contacted the *60 Minutes* film crew and asked them if they had photographs of the Liang Bua molds in Jacob's lab. Apparently yes, so Rex asked for copies. Damning evidence, especially if the photos were to be used in his review in *Nature*. But how this would all turn out was a worry. The damage to the holotype skeleton and the new hominid lower jaw did need to be officially documented by ARKENAS, for their own future protection—and for ours—but we seemed to be stirring up political forces and consequences way beyond our control.

By mid-March, Jacob and his team had submitted a brief paper to *Nature* titled "Large Errors in the Depiction of Small Humans," in which they correctly pointed out that the photograph of the femur in our paper describing LB1 was reversed, and the nearly complete femur was the left, rather than the right, as we described. None of their other criticisms had substance and some were misleading. For instance, they pointed to the fact that aside from LB1, only a single additional tooth was described in our paper. True enough, but they had spent days (or in Jacob's case months) examining all the borrowed Liang Bua hominid finds, which showed that there were at least 13 tiny individuals at the site with the same characteristics. They again concluded that LB1 was pathological and that

> [s]ince more than a year has elapsed between the discovery and description of these materials, we call upon the authors to provide a full, accurate presentation of all the evidence, as is their prerogative and obligation.

This was beyond irony. Our access to the Liang Bua hominid remains had been blocked for the past four months, which had delayed publication of our next paper, and some of the most important finds had been badly damaged while in the custody of the UGM paleoanthropology lab, so no one would ever be able to provide a full and accurate description of all the evidence. One of the referees said that the paper of Jacob et al. had no real substance, that they were playing a game of "gotcha," and that *Nature* or any other reputable journal should be above this type of behavior. The other referee objected strongly to the way that the material had been seized before the excavation team had a chance to study it, and said that the paper should be rejected—which it was.

Meanwhile, Harry Widianto had had an hour-long phone conversation with Dr. Etty Indriati, one of the UGM paleoanthropology lab assistants, who admitted that she, Nani Trilusiana Ratnawan and Janatin had taken molds of Liang Bua hominid remains—"but only of the teeth." She said that Koeshardjono was the person who had taken a mold of the second lower jaw, but would not say who had worked on the skull.

At Jacob's request, Harry then went to the paleoanthropology lab to discuss the increasing media interest in the story of the damaged bones. Harry had with him photographs I took at the Balai Arkeologi Jogyakarta offices immediately after the return of the hominid remains, showing the extensive damage to the second lower jaw. Jacob confirmed that molds had been made of the skull and two lower jaws, "but only low quality copies had been made before the moulds were destroyed"; that he was willing to return the remaining bones as soon as Tony Djubiantono came to Jogyakarta to collect them personally; and that the bones had been returned in the same condition as received—that any damage had been inflicted while still at the offices of ARKENAS in Jakarta.

The next day, Harry was recalled to the paleoanthropology lab by Jacob and given a letter that included demands that I apologize for slanders, such as claiming that Jacob had damaged the bones; explain to the Indonesian public whether the Liang Bua finds were owned by the Indonesian government or the researcher who provided the funds, and why Soejono had been replaced as the Indonesian chief investigator; and the time table and grand plan for research in Wallacea, Sulawesi and Java. He also requested that the bones not be housed at UNE; "not to change any evidences on the bones which are the bases of our conclusions about them"; and "do not take away the right of the Archaeological Centre to allow any anthropologist to study the bones." The letter concluded that

> [i]f you could conform to the conditions mentioned above and do not abuse in public Indonesian and other anthropologists who studied LB . . . then we the victims do not need to defend ourselves in the media, especially when you look at differences of scientific opinion with open mind.

The press was now really running with the story about the damaged Liang Bua bones. Articles appeared in the *Sydney Morning Herald*, *USA Today*, *Nature* and *Science*. Despite initial denials by Jacob that molds had not been taken, and insistence that no damage had been incurred in the UGM paleoanthropology lab, most reporters concluded otherwise, and *USA Today* also included some "before and after Jacob" photos of the second lower jaw in side view. In response, Jacob had three sets of photos showing the two Liang Bua lower jaws together in plan view sent to Dan Vergano, a reporter with *USA Today*: the first set had been taken at the end of December 2004; the second on February 19, 2005, a few days before the hand-back; and the third was taken by me a few days later, on February 23—the day of their return.

The difference in the condition of the second lower jaw in the first two sets of photos was striking. It started off almost identical in size to the LB1 lower jaw, but by February 19 the width of the dental arch and overall width had been reduced—the same condition we documented on February 23. The first two sets of photos taken at the paleoanthropology lab provided indisputable evidence for when and where the second lower jaw was damaged.

But there was good news as well. David Hamlin's film on the Flores research was first aired on the National Geographic Channel in America on March 13—and had been very well received—and John Rubin's film, *The Ultimate Survivor*, on early hominid sites around the world, including Liang Bua, was shown the following week.

There was also a press conference in Jakarta on March 28 to mark the first Indonesian edition of *National Geographic*, to come out in April. By happy chance, the cover story for 10 million copies of the magazine produced in nearly 30 different editions around the world just happened to be an Indonesian story—the finding of *Homo floresiensis* and its implications—with an article by Thomas Sutikna, Bert Roberts and Mike Morwood. The Indonesian president, Susilo Bambang Yudhoyono, senior people from the National Geographic Society in Washington and Asia, members of the press, researchers from ARKENAS and about 200 dignitaries attended the launch. It began at 7 p.m. at the Gedung Arsip Nasional with a children's choir, musicians, a banquet dinner and an archaeological display organized by ARKENAS. The display featured the LB1 lower jaw; skull and arm bones; the second, damaged lower jaw; a range of stone artifacts and animal bones; and in pride of place there was an anatomically accurate, lifelike model of the head of Hobbit.

Speeches were made emphasizing the importance of the new Indonesian magazine for promoting education, conservation and research—during which Terrance Adamson, the executive vice president of *National Geographic*, described *Homo floresiensis* as "the most astounding finding in world paleoanthropology in the last 50 years." After also speaking, President Yudhoyono signed a large poster of the first cover of the Indonesian version of the magazine featuring the depiction of a wild-eyed LB1, and was then presented with a framed photograph of the same. On the way out, President Yudhoyono and a large entourage visited the display to see LB1, guided by Thomas Sutikna, Rokus Awe Due and Harry Widianto. It was a great night for Indonesian archaeology.

But there was more in store. On April 3, the deputy minister for archaeology and history met with Tony Djubiantono, Soejono, Haris Sukendar and the Liang Bua excavation team to discuss a request from Soejono that all archaeology cooperations with foreign researchers should be reassessed. This meeting seemed to go in our favor, but then the Indonesian Academy of Science (LIPI) got in on the act, presumably with some prompting. They knew nothing about the 2001 agreement between ARKENAS and the University of New England, and decided to meet on April 14 with representatives from the Ministry of Culture and Tourism and the Ministry of Research and Technology to decide what to do. In the meantime, Tony sent me a fax suggesting that our research program be postponed until "everything becomes calm and clear."

Then Jean-Jacques Hublin resurfaced. With prompting, he wrote a letter to the rector of the University of Gadjah Mada complaining about how the Flores archaeological discoveries had been reported in the media and accusing us of launching a well-organized, efficient and aggressive media campaign against Jacob, in which his age, academic merit and ability to curate specimens were all questioned. It went on to criticize the exclusion of Soejono from the project by his ex-collaborators, noting that

> [a]lthough some almost invisible Indonesian archaeologists are still associated to the projects, a paleoanthropologist such as Prof T. Jacob represented a much more serious challenge.

Hublin concluded:

> I see this whole story as a pure example of scientific neo-colonialism, western arrogance and simply using people. It is also a lesson on the role of the media in our field and how cultural, linguistic and nationalist issues can interfere in a very negative way with science.

I thought that Hublin seemed to have a real aptitude for interpreting the activities of others in terms of his own behavior, while dismissing the younger researchers on the team as "some almost invisible Indonesian archaeologists" was insulting to them and their institutions. All of this was pointed out in my letter of response to the UGM rector. However, despite its many misleading or inaccurate statements, his letter was widely circulated amongst UGM staff, posted on their Web site, and forwarded to the Indonesian government unchallenged as part of an official complaint. Jacob was now calling for an official ban by LIPI on our Indonesian research. There was now another burst of publicity in the *Jakarta Post*, *Kompass* and overseas newspapers about a bizarre "Rampasasa Pygmy Somatology Expedition" to Flores in late April, led by Koeshardjono, that claimed to have found a village of 77 pygmy families close to Liang Bua. The team of six, including Jacob, had spent five days in the area measuring people, including two of our Liang Bua workers, Sius Sambut and Gaba Gaur, and concluded that, indeed, there were families of short people in the region. Another attempt to undermine the credibility of *Homo floresiensis* as a separate hominid species, but again no supporting evidence was produced.

More seriously, news about the LIPI meeting held on April 14 in Jakarta was not good. Representatives attended the meeting from the Ministries of Culture, Immigration and the Police, as well as Tony Djubiantono, Harry Truman Simanjuntak, Soejono and Jacob. In all, there were 14 people at the meeting, but no one from our excavation team: all of the "invisible Indonesian archaeologists" remained invisible.

What the end result of all this would be remained uncertain. Jacob was still sending letters of complaint to various ministries as well as LIPI. As a result, Tony's immediate superior, Hari Untoro Daradjat, deputy minister for culture, issued instructions that ARKENAS staff were no longer to publicly comment on the Liang Bua finds or the surrounding controversy.

Under Indonesian regulations, LIPI can terminate international cooperations without having to give reasons, and the message I got from Tony Djubiantono was that because LIPI thought that I had been conducting research in Indonesia without proper authority, it was "best to postpone the co-operation for a year." He went on to say that the real reason for all these problems was that our research had produced findings that were too spectacular.

So to clear things up, after talking with Tony and other staff at ARKENAS, I visited the main LIPI offices in Jakarta and discussed the situation with Ibu Kristiwati of the Bureau for Cooperation and Promotion of Science and Technology. The main problem apparently was that they had had no knowledge of the Agreement for Co-operation between UNE and ARKENAS, and had been led to believe that I was out there doing research without approval. Professor Dr. Umar Jenie, director of LIPI, was even quoted in the media as saying that I had been working illegally in Indonesia.

However, Kristiwati and later Dr. Neni Sintawardani, head of the bureau, seemed to accept the fact that my role in projects was now more one of coordination, advising and organizing finances, and that all specific projects were undertaken by Indonesian government research institutions and staff, who organized all the paperwork, permissions from local governments, personnel, letters of invitation for other participants, travel arrangements, purchases, finances and actual fieldwork. They concluded that I "had not been doing research illegally in Indonesia, but may not have met some of their regulatory requirements." I already had an Indonesian Business Visa (Research) for my role as project coordinator and institutional counterpart with the Geological Research and Development Centre, and the paleontological aspects of our research in Java, Flores and Timor continued to run relatively smoothly as did the associated training and exchange programs. However, getting formal approval from LIPI for further archaeological work at Liang Bua now required a letter from Tony Djubiantono saying that his institution also wanted to continue this collaboration.

During my next meeting with Tony about LIPI requirements and the need for another letter of support, he was clearly very worried about further controversy, more problems with Jacob and

Soejono, his institution's budget and his being replaced as director in the next restructuring. Most of all, as director of the Indonesian National Research Centre for Archaeology, he was worried that any further excavations on our part might yield further "spectacular findings, like another hominid skull" with all the attendant problems. It now seemed that our plans for further excavations at Liang Bua were to be postponed indefinitely.

EPILOGUE

Mid-2006 and it was good to be back in the Soa Basin. Fachroel Aziz and his GRDC colleague, Iwan Kurniawan, were huddled over a large, flaked stone artifact just exposed in a conglomerate beneath a layer of pink tuff near the base of the excavation. We already knew that the pink tuff was 880,000 years old, so this *in situ* artifact pushed back the time depth of hominids on Flores. But skeletal evidence for the hominids themselves remained elusive. A pity, because the whole point of our two-month-long excavation at Mata Menge was to get evidence for the type of hominids that initially colonized the island and were presumably ancestral to *Homo floresiensis*. Aziz and I reasoned that the greater the area excavated, the greater the chance of finding skeletal evidence for the associated hominids. We could speculate that the hominids that first reached Flores were probably small bodied and small brained with primitive body proportions, but as Dubois realized back in the 1880s, the bottom line for finding out what actually happened is the fossil record.

This will probably be the last major excavation we undertake in the Soa Basin. There are just too many other areas with potential to answer some of the fundamental questions posed when setting up the "Astride the Wallace Line" project. The Atambua Basin in Timor, for instance, is very similar to the Soa Basin with sites containing *Stegodon* fossils reportedly associated with stone artifacts. None of these sites has yet been investigated in detail or dated, so with hard work and some luck, I thought, there was a real chance of getting information that would be very useful in synthesizing the history of early animal and hominid in eastern Indonesia. The Kupang administration, which has administrative responsibility for Nusa Tenggara Timor, is also really keen to begin collaborative research in Timor to complement the Flores work. So keen that in June 2006, they organized a conference attended by more than 100 people at which Fachroel Aziz, Gert van den Bergh, Dik Dik Kosasih, Tony Djubiantono and I gave papers on the archaeology, paleontology and geology of Indonesia, with particular reference to Flores and Timor.

Other parts of Flores also beckon. While Aziz and colleagues were finishing the work at Mata Menge, an ARKENAS team, including Wahyu Saptomo, Thomas Sutikna, Atiek, Rokus and Wasisto, began excavations at Liang Panas, a large limestone cave in western Flores. This is the first of a number of excavations we planned to get evidence to compare with that from Liang Bua. At this stage, for instance, we don't even know if the dates for the extinction of *Homo floresiensis* and the arrival of modern humans are the same across the island. Possibly, modern humans first occupied coastal regions of Flores around the time they colonized Greater Australia, and only moved inland into the Liang Bua area much later. The answer lies in the soil.

Analysis of the Liang Bua evidence goes on, of course. Work on the *Homo floresiensis* material includes that of Bill Jungers and Susan G. Larson on the postcranials; Lorraine Cornish from the

Natural History Museum in London, who has been cleaning, repairing and stabilizing the bones; and Alan Cooper from the Australian Centre for Ancient DNA in Adelaide and Svante Pääbo from the Max Planck Institute in Germany, who are both attempting to extract DNA from small samples drilled from the pulp cavities of teeth. Specialist studies are also being undertaken on the excavated mollusk, rodent, bird, pig and *Stegodon* remains, with many more publications in the offing.

On related fronts, a recent independent study by Debbie Argue and colleagues featured a multivariate comparison between the cranial and postcranial traits of LB1 and those of modern people (including pygmies and microcephalics), *Homo ergaster*, *Homo erectus*, *Homo habilis*, one of the Dmanisi hominids, *Australopithecus* and *Paranthropus*. They concluded that LB1 was not a microcephalic human, cannot be attributed to any known species, and therefore is a new hominid species.

In contrast, a few critics continue to argue that LB1 is an abnormal, modern human pygmy. In August 2006, as this book was going to press, Jacob, Soejono, Indriati, Thorne, Eckhardt, Henneberg and others published a paper along these lines called "Pygmoid Australomelanesian *Homo sapiens* skeletal remains from Liang Bua, Flores: population affinities and pathological abnormalities." All the objections they raise in this paper are superficial and most can be countered with information presented in this book. For instance, they question whether Flores, as a relatively small island, could have supported a viable population of hominid hunter-gatherers for over 40,000 generations. Evidence for hominids in the Soa Basin spans some 200,000 years, however, meaning that long-term hominid occupation of the island was not only possible, but is well documented. Similarly, we have evidence for small-bodied hominids at Liang Bua from 95,000 to 12,000 years ago. How many years, or generations, does it take to demonstrate long-term population viability? Attempts by Jacob et al. to explain aspects of LB1 anatomy in terms of pathologies also ignore major traits that clearly differentiate LB1 from modern humans. For instance, the unique body proportions of *Homo floresiensis* are not like modern humans of any body size, including the smallest African pygmies and Andaman islanders. I believe that such arguments, based on cherry-picked evidence rather than the full range of *Homo floresiensis* traits and the context of the finds, are just not sustainable, and that, as in previous controversies about significant fossil hominid finds, maybe the issue is largely about conflicting ideologies and egos rather than science.

My main concern is in not countering such spoiler arguments, which later will only be of historic interest, but in trying to figure out the regional implications of *Homo floresiensis*, and how these can be used in planning future fieldwork programs. Of particular interest is Sulawesi, which has swept up, provided refuge for, and endemically shaped very early representatives of pigs, bovids, primates and marsupials, and which is the most likely source for the early hominids who reached Flores. Because of its strategic position, this island has enormous potential to shed light on the evolutionary and dispersal histories of extinct as well as living animal species. Were premodern hominids among the species that made it to Sulawesi? Almost certainly, but no one has really looked. I reckon that prospects for future archaeological and paleontological research on this most enigmatic of islands are very exciting, and planning is in train: during the excavations at Mata Menge this year, senior staff from Balai Arkeologi Makassar, which has responsibility for archaeological sites in southwest Sulawesi, paid a visit specifically to discuss possibilities for collaborative research with Aziz and me.

Going back one step further, the most likely source area for the land animals that colonized Sulawesi is the eastern edge of continental Asia, in the vicinity of the present-day island of Borneo—probably via the Philippines, with the island of Palawan being the most obvious transit route for animals migrating from the mainland to island Southeast Asia. During a recent visit to Tabon Cave, which overlooks the sea in southern Palawan, I was struck by its similarity to Liang Bua; by the fact that previous excavations at the site are extensive but relatively shallow; and that there are probably at least five meters' depth of unexcavated deposits at the site. I would bet money that the deposits at Tabon Cave contain evidence for premodern hominids, as well as for the arrival of modern humans.

Similarly, Niah Cave on Borneo has potential to provide archaeological evidence that greatly exceeds the currently known 50,000-year span for the site, and which may shed light on the ancestry of *Homo floresiensis*. Previous excavations in this huge cave went down a maximum of four meters, but its deposits could be tens of meters deep.

So far, the search for early hominids in Southeast Asia has concentrated on Java, where *Homo erectus* was in residence by 1.2 million years ago. However, the limited range of animal species present on Java before this time suggests it was a swampy, volcano-wracked landscape that may have made hominid colonization difficult. In other words, Java, like Flores, is not a good prospect for registering the earliest hominids in the region—unlike geologically older and more stable areas west of the Wallace Line, such as Borneo.

On a more general note, Indonesia has provided an important venue for the development of major scientific theories and disciplines. It was there in the 1850s that Alfred Russel Wallace initiated research into the factors determining the distribution of plants and animals. He thus founded the discipline now known as biogeography.

To Wallace, questions of species' origin and distribution were inextricably linked. And it was in Indonesia, while thrashing about in a malarial fever, that he conceived of natural selection as the driving force for the evolution of new species. The theory of evolution by Charles Darwin *and* Alfred Russel Wallace was first announced in 1858 at a Linnean Society meeting in London.

The first search for fossil evidence of the human past guided by a scientific theory—the theory of evolution—also took place in Indonesia in the 1880s. In his obsessive and groundbreaking search for fossil evidence of the "missing link" between apes and humans, Eugène Dubois not only found *Homo erectus*, but also founded the discipline of paleoanthropology.

These were all revolutionary developments in scientific thought, practice and discovery. Despite the continued efforts and finds of many other researchers—including Ralph von Koenigswald, Cornelius ter Haar, H. R. van Heekeren, Raden Pandji Soejono, Sartono, Teuku Jacob, Fachroel Aziz, Tony Djubiantono, Hisao Baba, Harry Truman Simanjuntak, François Semah and Harry Widianto—no further hominid species had been added to the two known from Dubois' time to have been present in Indonesia, *Homo erectus* and *Homo sapiens*. In the meantime, the focus for research on human origins and for revolutionary discoveries has moved elsewhere—to Africa, and for the last 40 years Indonesia, at the periphery of Eurasia, has also been considered peripheral to major developments in the story of human evolution. This entrenched way of thinking will change as more evidence emerges, and I agree wholeheartedly with Robin Dennell and Wil Roebroeks that "[m]ost probably, we are on the threshold of a profound transformation of our understanding of early hominin evolution." Evidence from Indonesia, and from Liang Bua specifically, will play a part in this transformation.

The discovery of Hobbit will therefore continue to impact generally on the discipline of paleoanthropology as well as personally on the lives and interests of all who participated in the Flores research. In fact, one of the most rewarding aspects of our project has been in subsequent career developments of mainly younger colleagues.

Thomas Sutikna has been offered a Ph.D. at the University of Wollongong on the prehistory of southwestern Sulawesi. He plans to excavate deeply stratified limestone rock-shelters on Sulawesi as part of his study of the cultural, faunal and (we hope) hominid sequence on this fascinating island.

Bert Roberts continues to date samples from Southeast Asian archaeological sites and obtained a large ARC grant to undertake similar work on early hominid sites in Africa.

Jatmiko is completing a master's thesis at the University of Indonesia for which he is carrying out excavations at the site of Kobatuwa in the Soa Basin of central Flores.

Carol Lentfer has obtained a large ARC grant to continue her work on landscape and cultivation histories in Melanesia and Southeast Asia, including analysis of residues on artifacts excavated from Liang Bua. Her work is providing information on long-term economic change at Liang Bua—in particular, when cultivation began.

Gert van den Bergh has obtained an ARC-funded research position at the University of Wollongong to continue work on the vertebrate paleontology of Southeast Asia.

Fachroel Aziz has been made a visiting professorial fellow at the University of Wollongong, and as a senior researcher is continuing to create research and educational opportunities for his younger GRDC colleagues.

Peter Brown continues with his analyses of the Liang Bua finds and lecture tours in the United States, Europe and Japan about *Homo floresiensis*.

Adam Brumm completed his Ph.D. on the technology of Soa Basin stone artifacts, and has applied for an ARC postdoctoral fellowship to extend his study of artifacts at early Asian sites.

Mark Moore completed his Ph.D. on Liang Bua stone artifacts and successfully applied for an ARC postdoctoral fellowship at the University of New England. He continues his research on the technology of stone artifacts in Australia, Southeast Asia, India and Africa.

Doug Hobbs, since his retirement from UNE, has begun a Ph.D. at the University of Wollongong on sourcing volcanic ash deposits in Java and Flores.

Kira Westaway completed her Ph.D. on landscape formation and paleoclimates in Flores and Java, and obtained an ARC post-doctoral fellowship at the University of Wollongong to continue her work in Southeast Asia.

Wahyu Saptomo is completing a master's thesis at the University of Indonesia for which he is carrying out excavations at the sites of Liang Michael and Liang Panas in western Flores. These limestone caves should provide much-needed information to compare with that obtained from Liang Bua.

Iwan Kurniawan is completing a diploma in paleontology at the Academy of Geology and Minerals in Bandung, Indonesia, and has

Some of the participants in the 2001 Liang Bua excavation. (PHOTO: MIKE MORWOOD)

undertaken Australian fieldwork and conservation training with the paleontology section of the Queensland Museum, Australia.

Kerrie Grant is presently doing forensic archaeology at Kurdish massacre sites in Iraq. When this work is completed she will continue her Ph.D. research at the University of New England on the ethnoarchaeology of pottery production in Flores, Java and Sulawesi.

Mike Morwood intends to continue research as long as he is able, and will maybe then be content to write and grow roses.

ACKNOWLEDGMENTS

The work described here was made possible by grants from the Australian Research Council, the University of New England and the University of Wollongong, as well as sponsorship from the National Geographic Society. Professor Michael Macklin (UNE Dean of Arts) and Dr. John Francis (*National Geographic*) helped greatly in obtaining additional financial backing. Craig Robinson of Melbourne made a private donation. Associated fieldwork was undertaken in collaboration with two Indonesian counterpart organizations, and in this regard we would like to thank Dr. Haris Sukendar, Dr. Tony Djubiantono and Professor Raden Pandji Soejono of the National Research Centre for Archaeology in Jakarta; and Mr. Bambang Dwiyanto, Dr. Djadjang Sukarna, Mr. Dikdik Kosasih and Dr. Fachroel Aziz of the Geological Research and Development Centre in Bandung.

Many other colleagues contributed their expertise to the projects, including Abraham Gampar, Adam Brumm, Bert Roberts, Bill Jungers, Carol Lentfer, Chris Turney, Doug Hobbs, Gert van den Bergh, Harry Truman Simanjuntak, Iwan Kurniawan, Jack Rink, Jacqueline Collins, Jatmiko, Jian-xin Zhao, Jose Abrantes, Kerrie Grant, Kira Westaway, Mangatas Situmorang, Mark

Moore, Michael Bird, Netty Polhaupessy, Paul O'Sullivan, Peter Brown, Rokus Awe Due, Sri Wasisto, Suminto, Susan G. Larson, Thomas Sutikna, Tular Sudarmadi, Victoria Paine, Wahyu Saptomo and Yani Yuiawati. We also acknowledge the generous support of local authorities, particularly staff from the Manggarai, West Manggarai, Ngadha and Nuse Tenggara Timor administrations, as well as local participants in the surveys and excavations, including Agus Mangga, Alex Gadhu, Andras Mali, Ansel Musa Ganda, Benyamin Tarus, Dius Nggaa, Domi Ben, Domi Deo, Ferri Bali, Flori Bali, Gaba Gaur, Ginus Denga, Kornelius Podha, Kristo Fores, Minggus Siga, the late Musa Bali, Peterus Mangar, Pit Ludu, Rikus Bandar, Rius Laru, Sius Sambut and Willem Lewa Nau.

Extensive editorial changes to previous drafts by Penny Jordan, Gert van den Bergh, Kathy Morwood, Mark Moore, Tim Whiting, Catherine Hill and Sara Foster (Random House) and Thomas Kelleher (Smithsonian) helped pull the book together, while Bert Roberts, Colin Groves, Dean Falk, Doug Hobbs, Fachroel Aziz, Iain Davidson, John de Vos and Thomas Sutikna checked sections. Doug Hobbs, Mike Roach and Kathy Morwood did line drawings with their usual style, while Penny Jordan also discussed ideas and formats for this book, and maintained an archive of publications, Web sites and associated e-mails, which has proved an enormously useful resource. We would also like to thank Bill Hamilton of A.M. Heath, our agent in London, and George Lucas from Inkwell, our American agent, for their negotiating efforts on our behalf.

Penny van Oosterzee would like to offer special thanks to a number of people for help with key insights into the sections on biogeography and sea currents. These include Robert Hall at the Royal Holloway Institute, London, for new information on plate tectonics and currents during glacial periods; Lars van den Hoeke-Ostende from the National Museum of National History, Leiden, for his expertise on rodents and their origins, for directing her to Antoni Alcover from the Institut Mediterrani d'Estudis Avancats, Mallorca, and for the pivotal information on the Balearic Islands and the sea currents surrounding them. Finally, Lawrence Heaney from the Field Museum, Chicago, enthusiastically provided valuable information on island biogeography, generally, and that relating to the Philippines in particular.

REFERENCES

The book has been written with readability in mind, and is not cluttered with citations and reference numbers. In keeping with this approach, key references have been provided below, divided into general areas of interest.

For more information on **paleoanthropology and insights into early human dispersal** refer to

Brunet, M., Alain Beauvilain, Yves Coppens, Emile Heintz, Aladji H. E. Moutaye and David Pilbeam (1995). "The first australopithecine 2,500 kilometres west of the Rift Valley (Chad)." *Nature* 378 (November 16):273–275.

Brunet, M., Franck Guy, David Pilbeam, Hassane Taisso Mackaye et al. (2002). "A new hominid from the Upper Miocene of Chad, Central Africa." *Nature* 418:145–151.

Dennell, R. (2004). "Hominid Dispersals and Asian Biogeography during the Lower and Early Middle Pleistocene c. 2.0–0.5 mya." *Asian Perspectives* 43:2.

Dennell, R., and Wil Roebroeks (2005). "An Asian perspective on early human dispersal from Africa." *Nature* 438 (December 22):1099–1104.

Finlayson, C. (2005). "Biogeography and evolution of the genus Homo." *Trends in Ecology and Evolution* 20(8):457–463.

Gabunia, L., Abesalom Vekua, David Lordkipanidze et al. (2000). "Earliest Pleistocene Hominid Cranial Remains from Dmanisi, Republic of Georgia: Taxonomy, Geological Setting, and Age." *Science* 288 (5468):1019–1025.

Gabunia, L., A. Vekua and D. Lordkipanidze (2000). "The environmental contexts of early human occupation of Georgia (Transcaucasia)." *Journal of Human Evolution* 38:785–802.

Johanson, D., and Maitland Eady (1981). *Lucy: The Beginnings of Humankind*, Touchstone, Simon & Schuster.

Kohn, Marek (2006). "Made in Savannahstan." *New Scientist* 1 July (2558):34–39.

Lindsay, E. H., Neil D. Opdyke and Noye M. Johnson (1980). "Pliocene dispersal of the horse Equus and late Cenozoic mammalian dispersal events." *Nature* 287 (September 11):135–138.

Mercer, John M., and V. Louise Roth (2003). "The effects of Cenozoic global change on squirrel phylogeny." *Science* 299 (5612):1568–1572.

Shen, G., Wei Wang, Qian Wang et al. (2002). "U-Series dating of Liujiang hominid site in Guangxi, Southern China." *Journal of Human Evolution* 43 (6):817–829.

Stringer, Chris, and Peter Andrews (2005). *The Complete World of Human Evolution*, Thames & Hudson.

Wolpoff, M. H. (1999). *Palaeoanthropology*, McGraw-Hill.

Wood, Bernard, and Mark Collard (1999). "The Human Genus." *Science* 284 (5411):65–71.

Key references on the **Wallace Line** and **plate tectonic movements in Southeast Asia** include:

Hall, R. (2001). "Cenozoic reconstructions of SE Asia and the SW Pacific: changing patterns of land and sea." In I. Metcalfe, Smith, J.M.B., Morwood, M. and Davidson, I.D., eds. *Faunal and Floral Migrations and Evolution in Southeast Asia-Australasia*. A.A. Balkema:35–56.

Hall, R. (2002). "Cenozoic Geological and Plate Tectonic Evolution of SE Asia and the SW Pacific: Computer-based reconstructions, model and animations." *Journal of Asian Earth Sciences* 20 (4):353–434.

Van Oosterzee, P. (1997). *Where Worlds Collide: The Wallace Line*, Reed Books.

Wallace, Alfred Russel (1883). *The Malay Archipelago: land of the orangutan and the bird of paradise*. Macmillan and Co.

For more information on **island biogeography** and **dwarfing** see:

Fleagle, John G. (1998). *Primate Adaptation and Evolution*. Second edition, Academic Press.

Heaney, L. R. (1985). "Zoogeographic evidence for middle and late Pleistocene land bridges to the Philippine Islands." *Modern Quaternary Research in Southeast Asia* 9:127–143.

Heaney, L. R. (1986). "Biogeography of mammals of Southeast Asia: Estimates of rates of colonization, extincition and speciation." *Biological Journal of the Linnean Society* 28:127–165.

Heaney, L. R. (2000). "Dynamic disequilibrium: a long-term, large-scale perpective on the equilibrium model of island biogeography." *Global Ecology & Biogeography* 9(1):59–74.

Heaney, L. R., Joseph S. Walsh Jr. and A. Townsend Peterson (2005). "The roles of geological history and colonization abilities in genetic differentiation between mammalian populations in the Philippine archipelago." *Journal of Biogeography* 32:229–247.

Jansa, S. A., Steven M. Goodman and Priscilla K. Tucker (1999). "Molecular phylogenetics of the native rodents of Madagascar: A test of the single origin hypothesis." *Cladistics* 15:253–270.

Kohler, M., and Salvador Moya-Sola (2006). "Reduction of brain and sense organs in the fossil insular bovid Myotragus." *Brain, Behaviour and Evolution* 63(3):125–140.

Leinders, J.J.M. (1984). "Hoplitomerycidae fam. nov. (Ruminantia, Mammalia) from Neogene fissure fillings in Gargano (Italy)." *Scripta Geologica* 70:1–68.

Lister, A. M. (1989). "Rapid dwarfing of red deer on Jersey in the Last Interglacial." *Nature* 342 (November 30):539–542.

Lomolino, M. V. (2005). "Body size evolution in insular vertebrates: generality of the island rule." *Journal of Biogeography* 32:1683–1699.

McFarlane, D., Ross D. E. MacPhee and Derek C. Ford (1998). "Body Size Variability and Sangamonian Extinction Model for Amblyrhiza, a West Indian Megafaunal Rodent." *Quaternary Research* 50:80–89.

Mittermeier, Russell A., et al. (1994). *Lemurs of Madagascar.* Conservation International Tropical Field Guide Series.

Palmer, M., G. X. Pons, I. Cambefort and J. A. Alcover (1999). "Historical processes and environmental factors as determinants of inter-island differences in endemic faunas: the case of the Balearic Islands." *Journal of Biogeography* 26:813–823.

Raia, Pasquale, Carmela Barbera and Maurizio Conte (2003). "The fast life of a dwarfed giant." *Evolutionary Ecology* 17(3):293–312.

Ruedi, M., Muriel Auberson, and Vincent Savolainen (1998). "Biogeography of Sulawesian shrews: Testing for their origin with a parametric bootstrap on molecular data." *Molecular Phylogenetics and Evolution* 9(3):567–571.

Schrago, Carlos G., and Claudia A.M. Russo (2003). "Timing the Origin of New World Monkeys." *Molecular Biology and Evolution* 20(10):1620–1625.

Schreiber, A. S., I. Seibold, G. Nötzold and M. Wink (1999). "Cytochrome *b* gene haplotypes characterize chromosomal lineages of anoa, the Sulawesi dwarf buffalo." *Journal of Heredity* 90 (1):165–176.

Schule, Wilhelm (1993). "Mammals, Vegetation and the Initial Human Settlement of the Mediterranean Islands: A Palaeoecological approach." *Journal of Biogeography* 20(4):399–411.

Van den Bergh, G. D. (1999). "The Late Neogene elephantoid-bearing faunas of Indonesia and their palaeozoogeographic implications: A study of the terrestrial faunal succession of Sulawesi, Flores and Java, including evidence for early hominid dispersal east of Wallace's Line." *Scripta Geologica* (117):1–419.

Van den Bergh, G. D., John de Vos and Paul Y. Sondaar (2001). "The Late Quaternary palaeogeography of mammal evolution in the Indonesian Archipelago." *Palaeogeography, Palaeoclimatology, Palaeoecology* 171:385–408.

Whitten, A. J., M. Mustafa, and G. S. Henderson, (1987) *The ecology of Sulawesi.* Gajah Mada Press. Jogyakarta, Indonesia

Information on **dating techniques** can be found in:

Aitken, M. J. (1999). "Archaeological dating using physical phenomena." *Reports on Progress in Physics* 62:99, 1333–1376.

Duller, G.A.T. (2001). "Dating methods: the role of geochronology in studies of human evolution and migration in southeast Asia and Australasia." *Progress in Physical Geography* 25 (2):267–276.

Roberts, R., Michael Bird, Hon Olley et al. (1998). "Optical and radiocarbon dating at Jinmium rock shelter in northern Australia." *Nature* 393 6683(May 28):358–362.

Roberts, R. G., M. J. Morwood and Kira E. Westaway (2005). "Illuminating southeast Asian prehistory: New archaeological and paleoanthropological frontiers for luminescence dating." *Asian Perspectives* 44(2):293–319.

For information on **Flores archaeology and the Soa Basin** see:

O'Sullivan, Paul B., Mike Morwood, Douglas Hobbs, Fachroel Aziz Suminto, Mangatas Situmorang, Asaf Raza and Roland Maas. "Archaeological implications of the geology and chronology of the Sea Basin, Flores, Indonesia." *Geology* 29(7):607–610.

References for **Ngadha and Manggarai culture** include:

Erb, M. (1987). When rocks were young and earth was soft: ritual and mythology in northeastern Manggarai. *Anthropology*, State University of New York.

Erb, M. (1999). *The Manggarians.* Times Editions, Singapore.

Java's fossil record:

Kaifu Yousuke, H. B., Fachroel Aziz, Etty Indriati et al. (2005). "Taxonomic affinities and evolutionary history of the early pleistocene hominids of Java: Dentognathic Evidence." *American Journal of Physical Anthropology* 128(4):709–726.

Larrick, Roy, Russell L. Ciochon, Yahdi Zaim et al. (2001). "Early Pleistocene ^{40}Ar/^{39}Ar ages for Bapang Formation hominins, Central Jawa, Indonesia" *PNAS* 98(9):4866–4871.

Modern human evolution and dispersal:

Kuhnt, W., Ann Holbourn, Robert Hall et al. (2004). "Neogene history of the Indonesian throughflow." *Continent-Ocean Interactions within East Asian Marginal Seas* Geographical Monograph Series 149(3):299–321.

Macaulay, V., Catherine Hill, Alessandro Achilli et al. (2005). "Single, Rapid Coastal Settlement of Asia Revealed by Analysis of Complete Mitochondrial Genomes." *Science* 308(5724):1034–1036.

Schule, Wilhelm (1993). "Mammals, Vegetation and the Initial Human Settlement of the Mediterranean Islands: A Palaeoecological Approach." *Journal of Biogeography* 20(4):399–411.

Thangaraj, Kumarasamy, Gyaneshwer Chaubey, Toomas Kivisild et al. (2005). "Reconstructing the Origin of Andaman Islanders." *Science* 308(5724):996

References for **Australian archaeology** include:

Bowler, J. M., Harvey Johnston, Jon M. Olley et al. (2003). "New Ages for human occupation and climatic change at Lake Mungo, Australia." *Nature* 421 (February 20):837–840.

O'Connell, J. F., and J. Allen (2004). "Dating the colonization of Sahul (Pleistocene Australia–New Guinea): a review of recent research." *Journal of Archaeological Science* 31:835–853.

Birdsell, J. H. (1977). "The recalibration of a paradigm for the first peopling of Greater Australia." In *Sunda and Sahul*, eds. J. Allen *et al.*, pp 113–168. Academic Press, London.

Walsh, O. L. (2000). *Bradshaw Art of the Kimberly.* Takarakka Nowen Kas Publications, Towong, Queensland.

The **history of cultivation, domestication and the Austronesians** is covered in:

Bellwood, P. (1997). *Prehistory of the Indo-Malaysian Archipelago.* University of Hawaii Press.

Bellwood, P. (2001). "Early Agriculturalist Population Diasporas? Farming, Languages and Genes." *Annual Review of Anthropology* 30:181–207.

Bellwood, P. (2005). *First Farmers: The Origins of Agricultural Societies*, Blackwell Publishing.

Larson, G., Keith Dobney, Umberto Albarella et al. (2005). "Worldwide Phylogeography of Wild Boar Reveals Multiple Centers of Pig Domestication." *Science* 307(5715):1618–1621.

For detailed information on ***Homo floresiensis*** **morphology** see:

Brown, P., T. Sutikna, M. J. Morwood et al. (2004). "A new small-bodied hominin from the Late Pleistocene of Flores, Indonesia." *Nature* 431 (October 28):1055–1061.

Falk, D., Charles Hildebolt, Kirk Smith et al. (2005). "The Brain of LB1, *Homo floresiensis*." *Science* 308(5719):242–245.

Jacob, T., E. Indriati, R. P. Soejono et al. (2006) "Pygmoid Australomelanesian *Homo sapiens* skeletal remains from Liang Bua, Flores: Population affinities and pathological abnormalities." *PNAS* 102(36) 13421–6.

Morwood, M. J.,v R. P. Soejono, R. G. Roberts et al. (2004). "Archaeology and age of a new hominin from Flores in eastern Indonesia." *Nature* 431 (October 28):1087–1091.

For information on **Stone Age technology** see:

Brumm, A., F. Aziz, G. D. van den Berg et al. (2006). "Early stone technology on Flores and its implications for *Homo floresiensis*." *Nature* 441 (June 1):624–628.

Media reaction to the discovery of ***Homo floresiensis*** includes the following:

Diamond, J. (2004). "The Astonishing Micropygmies." *Science* 306 (5704):2047–2048.